McGUINNESS

Interpreting the art of Tom McGuinness

Robert McManners and Gillian Wales

By the same authors:-

Tom McGuinness - The art of an underground miner

Memories of Witton Park

Shafts of Light - Mining art in the Great Northern Coalfield

Published by Gemini Productions

ISBN: 0-9532217-2-5

Designed and printed by Hillprint Media 01325 245555

Dedicated to The Bevin Boys -
especially Tom McGuinness

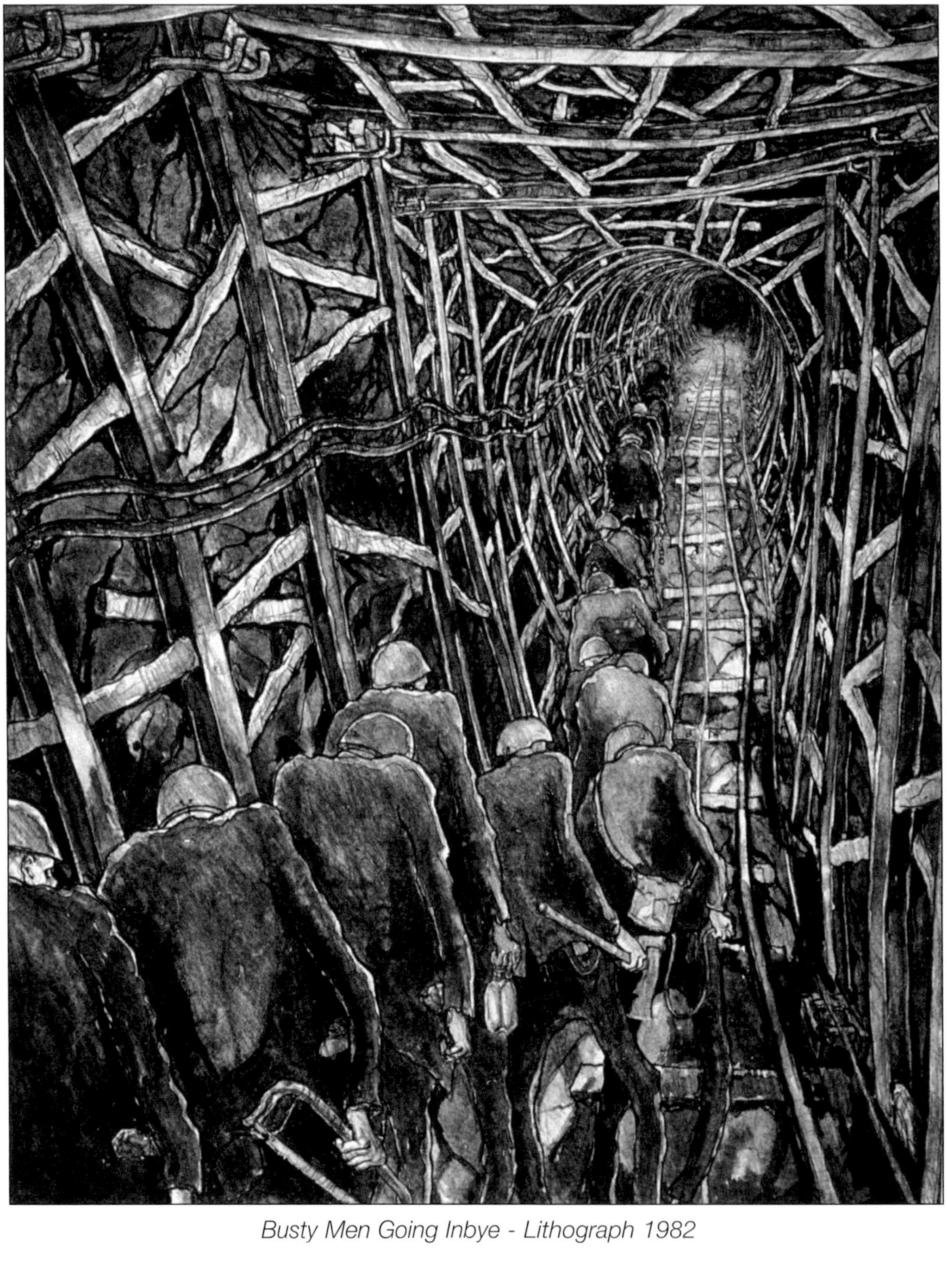

Busty Men Going Inbye - Lithograph 1982

Contents

Acknowledgements

As authors of this book we would like to express our extreme gratitude to all those who have assisted in its production.

During the past ten years we have held conversations with and have interviewed many of Tom McGuinness's friends, colleagues and family members. Without exception each has given time and information freely and enthusiastically which has made our task not only easier but infinitely more satisfying. Whilst we extend our thanks to all those individuals there are many that we feel we must thank specifically.

In this respect we would like to record our gratitude to Rosemary Davidson, Gill Holloway and Roy Thomas. There are those who sadly are no longer alive to receive our gratitude but whose help has been invaluable to us including Winnie Allison, Nerys Johnson, Sid Lockey and Jack Reading. We would like to acknowledge the assistance we have received over the years from Tom's immediate and extended family, particuarly sons Shaun and Anthony and daughter Corinne.

The high quality of the photography for the book is due to the expertise and patience of photographer Syd Neville. We would equally like to thank Elaine Kenyon and George Teasdale for scanning and adjusting the images, Lintons Printers for their integrity and Hillprint for their expertise throughout the production process. Thanks are also due to our proof-readers, Elaine Kenyon, Aileen Little, Stefa McManners, Wendy Robertson and Janet Strong and to the galleries and owners of paintings in allowing us to reproduce the art works in their ownership.

Special thanks are due to Elizabeth Conran for her advice and for her perceptive foreword to the book and to Bernard Hathaway for his generous sponsorship.

Sincere thanks are due to our respective spouses, Stefa and Glynn for the patience, encouragement and enthusiasm they have shown us throughout this project.

Our deepest gratitude is, however, reserved for Tom McGuinness who has patiently advised us and explained to us the art from his own perspective. We have come to value his friendship as we do his art.

Foreword

I have had some resistance in the past to 'mining art', stemming from the rather uncritical attitude which surrounds it. There's a very loyal northern view that if it's a mining subject, it must be worthy, compounded by a 'post Ashington Group' view, that all mining art is produced by talented Sunday painter miners, and so, even worthier still. In the context of national and international art, it makes mining art look like a poor provincial backwater.

Robert McManners and Gillian Wales have overturned these views. They have collected fresh information from many miner-artists and their families, as well as a huge number of images of works of art. They have shown perceptive appreciation of this form of art, rather than uncritical adulation. Thanks to their work, mining art can be seen as a serious regional art movement, and Tom McGuinness as one of its brightest stars.

Tom McGuinness is less a miner who happens to be an artist, than a dedicated artist who has supported his family by working as a coal-miner. He has sought instruction all his life to develop his artistic skills and techniques. His art is not socialist-heroic or couthy narrative. It is, most surprisingly, lyrical and poetic, resembling scenes from a ballet.

In his work, the interiors of the mines are like stage sets. He explores their vertical shafts, horizontal tunnels, stairways, forests of pit props, and spiders' webs of tunnel girders. His compositions are constantly inventive and varied. His figures, like dancers, respond to the spaces, pressing against walls, squeezing into seams, moving in hunched ensembles along tunnels and up and down stairs. They are silent, vulnerable, lonely. They live with a constant awareness of death. And for a subject which exists underground, overground light plays a major role - white moonlight, the gentle serene, precise light of the full moon.

One of the secrets of Tom McGuinness's success is his considerable knowledge of art history, culled mostly from books. This knowledge has been totally absorbed into his own work, and transformed, so that it is very difficult to pinpoint which artist or art works he has drawn on. For me, there are echoes of Piranesi, Fritz Lang's 'Metropolis' and, dare I say it, post-World War Two School of Paris abstraction. This ingredient gives his mining art a special sophistication and universal attractiveness. It makes him an artist of far more than regional importance.

Elizabeth Conran
Art Historian
2006

Preface

In 1997, we wrote and published 'Tom McGuinness - The Art of an Underground Miner'. This was an in-depth biographical study of the major events in Tom's life and how those events informed the basis of much of his art.

Tom McGuinness will be eighty in 2006. He still paints daily. This book is a tribute, not only to those eighty years, but is an attempt to contextualise and give an insight into the art of this unique man.

Not a biographical account, the book focuses on the influences, the constraints and the motivations behind his art. We consider the impact made by individuals, by other artistic influences and by life events. For over ten years we have had the privilege of meeting with Tom regularly, discussing his pictures, his philosophy and the interpretation of his art. In the main body of the text we give an insight into Tom's imagery - the style, the subject, the technique, the composition, but especially the meaning. With Tom's advice we have analysed selected pivotal works which illuminate these elements. In the light of this our personal commentary accompanies each image. The viewer interprets art from the standpoint of his or her own culture, societal status from within that culture and from the perspective of the values held.

It is easy to recognise Tom's work as a social and industrial documentary and to appreciate his craft for its expertise and finesse and for his ability to manipulate the images - his use of line, tone, colour and distortion. The interpretation and the meaning behind his images are not so easily accessed. It is, however, that very iconology that raises Tom's work to the level of great art. Tom has always, rightly, had confidence in his own artistic ability and we try to analyse the motivational influences that fuelled the determination and the drive which encouraged a Durham miner from humble origins to emerge as one of Britain's most powerful and significant industrial artists.

Influences came from many sources. Individuals positively encouraged Tom, notably his family, his boyhood teacher and his priest. Later, art tutors and fellow artists he came into contact with at various institutions had impact on his artistic direction, especially in his formative years. Chance encounters and major life events can be traced by their influence on his art. External sources were, however, not the main motivational or driving forces: these had to come from within.

Tom himself has actively sought knowledge in all facets of art: the practical, the theoretical, the historical, the philosophical and the iconological. He has a vast insight and awareness of all aspects of the art world, largely gained through his avid reading and his own self education.

Tom, a modest, quiet man, judges success by the critical acclaim of those who truly value his work - not those who collect it as a potential investment, but those who treasure his paintings for the statements they make. Whether this acclaim be from the miner who buys an etching for nostalgia or the considered evaluation of a fellow artist or critic is immaterial to Tom. It is the sincerity of the appraisal that he appreciates and values.

It is no accident that Tom's trademark style and imagery are as we now see them. They are the result of a long process of artistic evolution based on personal experience and Tom's fascination for his fellow man.

Robert McManners and Gillian Wales - January 2006

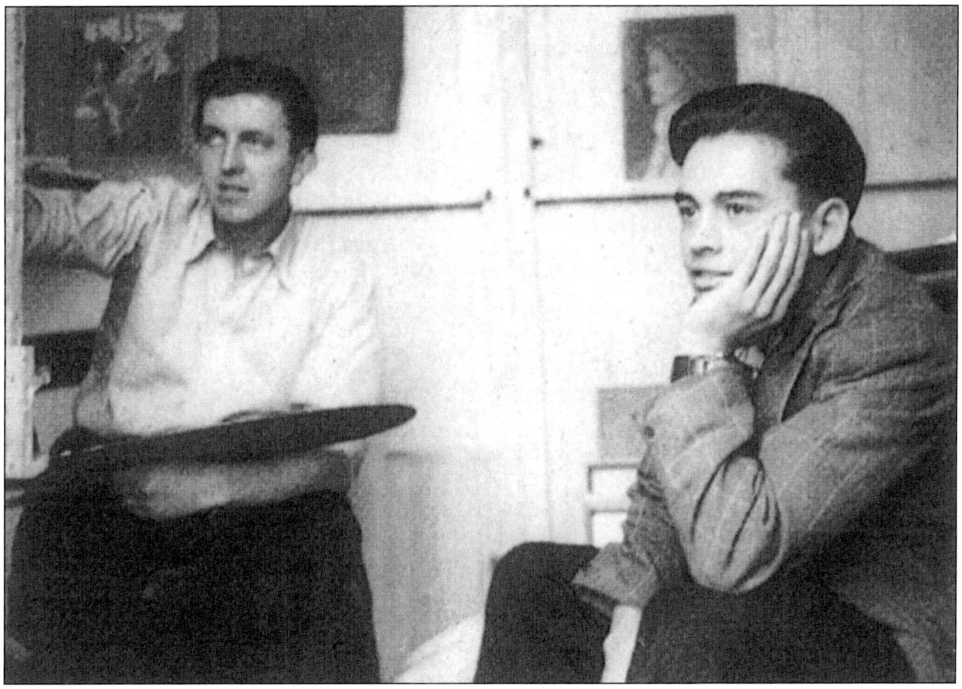

Tom painting in his studio - his shed - in the 1950s

Introduction

We each became independently aware of the work of Tom McGuinness in the early 1970s. The life of the miner suddenly became real as Tom's dark meticulous etchings exposed that arcane world to us. From that time we have been compelled to understand the subtle strength and unique iconology of this most unassuming of artists. Although we were both raised in active mining communities it was the emotive power inherent in Tom's pictures that revealed to us the stark reality of the underground experience.

Tom's work undoubtedly epitomises and is firmly rooted in social realism - his body of work truly concerns itself with the social impact of the mining industry - yet the depth of emotion revealed raises Tom's work above that of the documentary or the mere vernacular and places him at the forefront of contemporary expressionism.

Rosemary Davidson, owner of the Broughton House Gallery, Cambridge, supports this notion in asserting, "To categorise McGuinness's work as social history painting would be quite wrong. He does indeed record the life of the mining communities that have ceased to exist as such ... bending, crawling, standing, lying, standing half naked in the underground galleries, the men suggest figures from Dante's Inferno and the waiting women those of Greek tragedy. It is this capacity for tapping the universal in human experience that sets the work of McGuinness apart from that of other artists of the industrial scene. His styles put one in mind of German Expressionists such as Kathe Kollwitz."

The term expressionism describes art in which conventional ideas of realism and proportion seem to have been overridden by the artist's emotion, with resultant distortion of shape and colour. Tom's work meets these criteria in that his strong use of colour and his distortion of the image, almost to the point of abstraction, create a sense of conflict against the elements and society and represent his search for truth - a truth which requires him to break with tradition and portray the life of the miner in all its unacceptable reality. A confining, claustrophobic, clamorous reality of dark and filthy conditions - conditions in which working miners were often unable to stand upright - frequently robbing the miner of his health and even his life.

Jack Reading, Welfare Officer of the Coal Industry Social Welfare Organisation and a major collector of Tom's work says, "Tom may seem to caricature and purposely distort to the point where legs become pit props and arms become indistinguishable from the arched girders which contain them, but he is a privileged observer. What he sees he has felt; what he depicts he has experienced; and that is rare in an artist."

Tom's art results from his personal response to the reality of life underground. Whilst embracing the legitimate technique of distortion to intensify the emotive power of the picture, his work retains a social realism - a striving to record the world as he sees it - at times ignoring the conventions, to focus on the subject matter and its emotional and social impact. "An early tendency to stylise figures as flexible, elongated shapes has now ceased to be a mannerism and become a strength of design, permitting him to impose haunting interior rhythms in his placing of these bowed, malleable forms with a skeletal backdrop of props and cages," writes art critic Gail-Nina Anderson.

All this is not to undervalue the importance of Tom's work in recording the way of life of the miner for future generations - a way of life now consigned to history. Whilst Tom's images are a personal record of mining life, above all Tom paints as a true artist - for himself - with an insight, an honesty and an integrity which are the hallmarks of experiential art. His desire to interpret the miner's lot has inspired Tom, a working pitman, to produce the best, most accurate and complete artistic body of work of any industry.

Tom McGuinness's work withstands critical scrutiny in its own right, for its technical expertise, for its personal style and for the fascination of its subject matter. It is all the more remarkable as it represents the personal diary of an underground miner. Tom's work is his autobiography. It has been our privilege to have spent so much time in the company of this most considerate and unassuming man, discussing his work and its intrinsic philosophy. This book is both a tribute to an important artist and an interpretation of his unique body of work.

"I have always tried to express the feeling of restlessness in my work ... Underground the roof lowers, the girders bend, the miner stoops and the cycle continues."

Tom McGuinness

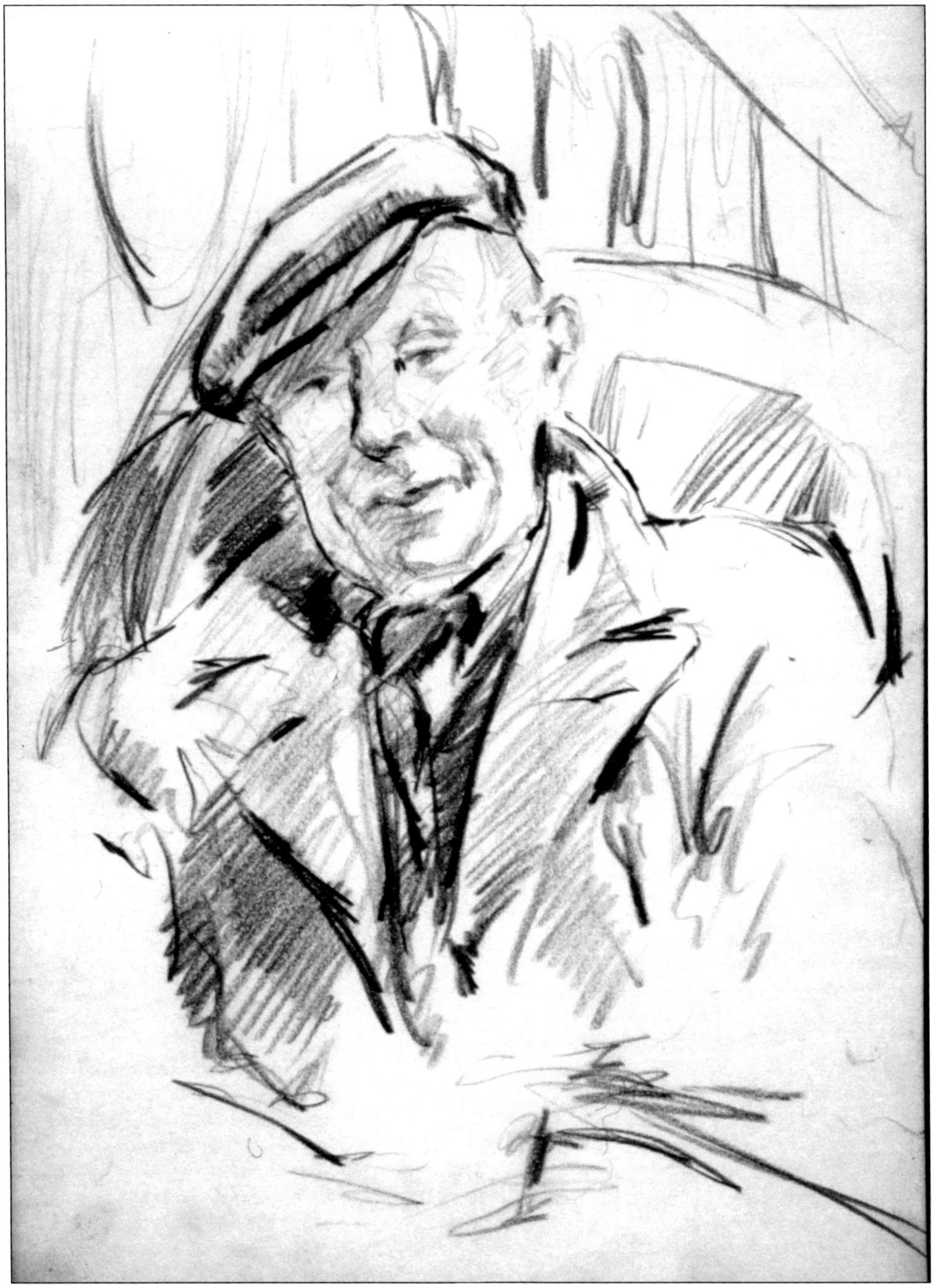

The Smiling Miner - Graphite and Charcoal 1949

Influences on the Art of Tom McGuinness

Amongst the heavy industries that dominated the North East's economy in the nineteenth and twentieth centuries, coalmining was unique. Not only was it the powerhouse of the Industrial Revolution, spawning other industries, it also brought with it a complete way of life for those involved. The coal industry created its own communities, its own language and customs - indeed its own culture. Whilst this pattern may have occured to a smaller degree in other heavy industries nowhere do we find a culture whose sophistication includes a corpus of experiential art. Not only did the miners paint the mines but there were elements in the fascination of the coal industry which attracted the attention of the professional artist. The spoken language of the North Eastern miner may have been 'Pitmatic' but it seems that there was also a need for the miner to communicate in a different language - the language of art and imagery. There were many exponents of this language - some with a greater pictorial vocabulary and command of the image than others. No one used this form of communication as skilfully and as eruditely, with as many nuances as Tom McGuinness. This form of communication is becoming lost to us which makes the imperative of understanding and appreciating its greatest exponent the more compelling.

Throughout his working life, Tom McGuinness, by means of his art, documented the miner's world from direct experience. From his early days as a collier, Tom, the artist, took the observer straight down the mine. His work is contemporaneous, many of his sketches being drawn underground. He has filled a vast number of sketch-books over the years. Each page is covered in a myriad of small drawings, almost mosaic-like, in which he has recorded images observed whilst at work. These initial observations would be developed into artistic ideas which became the inspiration for future paintings. Where sketching was impossible Tom used plasticine to model figures of men working in the low seams which, in their turn, were incorporated into drawings and paintings.

In this aspect of his work he is unique, as almost all other major bodies of underground mining scenes, including those by Ted Holloway, Oliver Kilbourn and Bob Olly, are retrospective. Even commissioned war artist Henry Moore remembered the coalface workers from his two weeks of exposure to coal mining at Castleford and 'drew them up' over the subsequent two months. In this, Tom's body of work can never be repeated or rivalled. There cannot ever again be an artist who will spend forty years working through times of staggering industrial and technological advances recording the conditions and emotions of the miner with such technical finesse and with such an incisive eye as has Tom McGuinness. He could not have envisaged that his lifespan as a mining artist would extend beyond the remaining lifespan of the Durham Coalfield and that in his later days Tom would be recording the demise and social aftermath of a once great and seemingly indestructible industry.

Tom was born in 1926, the year of the General Strike and the protracted Miners' Strike, into what had officially been described in Parliament as County Durham's "most distressed village" - Witton Park. This, like many other North East towns, had been entirely a product and a victim of the Industrial Revolution. It owed its existence to the chance proximity of all the mineral elements required for iron making but particularly the readily available high quality coals needed for the smelting process. The rapid growth of the town started in 1846 with the opening of the Bolckow and Vaughan Iron Company's works at the incongruously named Paradise Fields. Within twenty five years Witton Park's population rose from 91 to over 4000. To accommodate their 1800, mostly immigrant, workers Bolckow and Vaughan hastily built poor quality two up two down random rubble terraces in a gridiron pattern to maximise the density of habitation. The boom was not to last and the popularity of steel meant a drastic fall in the demand for Witton Park's pig-iron. The works closed in 1884 with the effect of bringing closure to almost all of the surrounding collieries and leading to a male unemployment rate of 96% and utter and abject poverty. Even the 'relief' of The Great War, when the men from Witton Park volunteered in the highest ratio nationally, was not sustained and Witton Park endured its Great Depression long before that malaise caught up with rest of the country in the 1930s.

It was into such deprivation that the young Tom McGuinness was thrust. Jobs were at a premium and after leaving the local Roman Catholic school, St Chad's, in 1940, he took any job he could find until 1944 when he was conscripted to the coal industry as a Bevin Boy. In 1943 Ernest Bevin, the Minister of Labour and National Service, under the coalition government's emergency powers, resorted to compulsory conscription into the mining industry because of the falling coal production due to an ageing and dwindling workforce. Those who were to serve in the coalmines, as opposed to the armed forces, would be selected by impartial ballot - hence the term 'Bevin Boy'. Thus what was to prove to be Tom McGuinness's life-long obsession with coal and the mining industry happened as a result of sheer chance. Tom was to remain in mining for thirty nine years until he was made redundant in 1983, having worked in many of Durham County's collieries and having experienced many different working conditions and practices.

Tom concedes that, although he was in the Home Guard at Witton Park, the colliery shift system meant the Second World War did not impact significantly on him, other than that he rose to the added responsibilty of being the 'man of the family', looking after his Aunt Winnie and her sister Nellie with whom he then lived, whilst his uncle was away in the war. Tom had harboured no ambition to become a collier as there weren't any deep mines surviving in the Witton Park area - his conscription meant him working further afield than his previous jobs had demanded. Had he not become a collier Tom feels that his art may not have held the same importance for him. He is unsure that he would have found the powerful artistic inspiration in other industries in which he had worked.

Witton Park was to prove both inspirational and formative for the evolving artist Tom McGuinness. More interested in the outdoors than school, days he spent by the River Wear observing the derelict landscape about him stimulated his powers of observation. He became fascinated by the intricacies of abandoned engine houses, obsolete pithead gear and dilapidated machinery. His emerging observational faculties would be one of the fundamental skills that would underpin Tom's art. In future years he would find inspiration in and be fascinated by everyday human behaviour and memorise scenes from his daily work. "There was always something interesting happening and to be observed on the way to work - from men with flat caps, to posters on walls, to pigeon racing ..." 'Early Morning Shift', (Plate 10) a painting that Tom considers to be one of his best, epitomises those powers of observation. Inspiration struck while travelling on the early train past the Black Boy Colliery, Eldon. Observing the scene from the raised vantage point of the railway embankment, looking down on the streets, the idea for the painting was conceived. "Draw what you see, draw while fresh in the mind, work at it, improve, change," Tom affirms.

He had always been interested in drawing at school but had had no formal training until his Colliery Training Officer, intrigued and impressed on finding Tom drawing in chalk on the side of a coal tub, encouraged him to attend evening classes. He enrolled at Darlington School of Art in 1944 and attended for six years, despite the difficulties of rearranging shift work and the uncertainty of wartime transport. He studied life drawing, portraiture, and 'antique drawing' from casts. His original art teacher, Ralph Swinden, Tom tells us, significantly influenced his style by encouraging him to abandon his tentative approach to drawing and to adopt a much bolder line. These classes did not provide formal teaching in anatomy and so Tom taught himself from Dunlop's 'Anatomy for Art Students' and from anatomical charts - such was his determination for accuracy and understanding of the principles of human movement and structure.

Tom's interest in art prompted him to travel to see national touring exhibitions including the work of Turner in the Laing Art Gallery, Newcastle and Daumier in both Glasgow and the Victoria and Albert Museum in London. These artists were to greatly influence Tom - the lithographs of Daumier, in particular, were a source of lasting inspiration.

Daumier's influence can be seen in the physical size of the characters in Tom's lithographs who are generally portrayed much larger, are more exaggerated in their movement and are delineated in a more flowing style - a freedom afforded by the medium of lithography - than those who inhabit his etchings and engravings. The intaglio print techniques of etching, aquatint and engraving - in which the plate surface is scored - does, however, afford Tom a different opportunity for expression. Tom uses the precise nature of the image from the etched or engraved plate to collude with the observer in creating an impression of confinement and claustrophobia, which offers the viewer a feeling of intimacy that implies the sharing of his arcane world. He became an avid reader and collector of art literature, especially of the etchings of Rembrandt, Durer and Goya. Elements of the etching styles of all three artists are apparent in his work.

In common with the rest of the Bevin Boys, Tom was still working in the mines in 1947, two years after the war had ended. This, coupled with the government's decision not to offer the Bevin Boys any financial or other material recognition for their years of enforced hardship, left many of them bitter and determined to quit mining forever. Following the nationalisation of the coalmines on Vesting Day, 1st January 1947, the National Coal Board appealed to the miners to remain in the industry because the post-war need for coal was greater than ever. Although some of the Bevin Boys remained, most left, as did Tom McGuinness who went to work briefly at various jobs in other industries - but the lure of the mines was too strong.

In 1948, Tom joined The Spennymoor Settlement and enrolled in the Sketching Club. The Spennymoor Settlement was an educational and recreational organisation which had been founded in 1931 by the Pilgrim Trust "to attempt to alleviate some of the social evils arising from the economic disaster which hit County Durham ... worse than any other of the industrial areas," wrote Bill Farrell, its first warden. It was here that Tom met the writer Sid Chaplin and other pitman artists, including Norman Cornish. Tom and Norman recognised each other as kindred spirits and they became firm friends.

The influence of the inspirational Warden Bill Farrell and The Spennymoor Settlement had encouraged Norman Cornish in his life-long passion for art and had given him an artistic direction when he had joined in the 1930s. Tom McGuinness was similarly inspired when he joined in the late 1940s. Norman had a much more singular view of his art than did Tom. Both men, however, heeded Bill Farrell's advice, 'to paint what they knew'. Bill Farrell would not allow copying and all illustrative work at The Spennymoor Settlement Sketching Club had to be drawn from life. Bill was promoting experiential art and this simple expedient can be observed throughout Tom's work.

Tom took Bill's advice literally and his first example was a drawing of a young man with a badly injured arm; a scene from Blackett's Saw Mill where Tom was then working.

This was to be the first time Tom used his own life model. Prior to this neither his work mates nor his working environment had featured in his art. He was being encouraged to sketch from his own experience. Of equal importance to Bill Farrell was the ethos of using the best materials; a principle to which Tom has always adhered.

The Sketching Club at The Settlement was not a class in that there was no art tuition. Education was by mutual problem solving, by asking questions about technique, by debate, argument and criticism of style - similar to the teaching techniques that Robert Lyon was later to adopt for the Ashington Group of artists. Tom profited greatly from his experience at The Settlement for it was at this time that he began to paint seriously, mainly portraiture. Local portrait commissions and invitations to exhibit followed the purchase in 1949 by the Shipley Art Gallery, Gateshead, of the conté drawing 'Miner and Child'. (Plate 1)

Realising that he needed a dedicated studio Tom began renting an attic room in a theatrical boarding house in Princes Street, Bishop Auckland where he could work without fear of interruption. Jack Reading in his foreword to 'Tom McGuinness - The Art of an Underground Miner' commented, "I was impressed even from the start of our acquaintance by the dedication to his art by always contriving to have, like a true professional, a place set aside for a studio - be it a garden shed or a rented room nearby - and by his insistence on using materials of the best quality." Tom continued to attend The Settlement until he returned to art classes in Darlington in 1951, exhibiting annually with the Darlington Art Society. At this time he produced his first oil paintings.

In 1956 he met Gill Harman (later Holloway), art tutor at Durham University Extra-Mural Department, who was to be a major influence. She taught many of the men from the North East of England who were to go on to become professional painters, including Ken Cozens (later Curator of Middlesbrough Art Gallery) and Ted Holloway (later Head of Art at Jarrow Grammar School and to become Gill's husband). Gill instituted a revolutionary new method of teaching art to adults. She believed strongly that the practice of art must be taught in parallel with the study of art history and the major works of art. Only by learning in this way was a student given both the intellectual and the practical skills needed to become an artist of note. These extra-mural classes acted as a magnet for those who had missed out on education first time around - steel workers from Consett, men from the shipbuilding industry, together with many regular miners. Of Tom, Gill later observed, "He is a man of few words yet he is a man of great strength with much to say. What he says he says with paint ... What marks him as an artist of rare merit is his ability to open our eyes to the beauty that exists even within such subjects ... the blacks, browns and ochres of his early paintings of the 50s soon gave way to those in which translucent glazes of brilliant blues and greens predominate ... thus the darkness of the underground scene is transformed by these rich, transparent colours. Light reflects from the canvas illuminating the whole and resulting in works which contain a very rare beauty."

Tom had staged his first solo exhibition in 1958 in the offices of the Coal Industry Social Welfare Organisation at Hobart House in London, organised by their Welfare Officer, Jack Reading. The lack of an art gallery in Durham City was resolved when Gill, realising how vital it was to rectify this omission, persuaded the owner of the City Hotel to dedicate a room as a gallery. Here exhibitions of the work of many contemporary British artists were held and the hotel became a meeting place for artists, critics, art historians and patrons. Tom exhibited here, meeting many other artists and exchanging ideas.

The public's opportunity to compare the local pitmen artists Tom McGuinness and Norman Cornish with the nationally-known Josef Herman - a Polish emigré who had illustrated the Welsh coalfield and who painted in an expressionist manner - came in November 1969 when the three artists exhibited at the Stone Gallery in St Mary's Place, Newcastle. Bill Johnson, art critic of the Manchester Guardian, in his review of 30th November concluded, "McGuinness is at the top of his form in two charcoal portraits of miners and Cornish continues in his own humanitarian way with pastel figure studies akin to Degas. Josef Herman, in a series of pen and wash drawings in which monumentality is suggested by a four-square ruggedness, is possibly not at his best. Certainly he suffers by comparison."

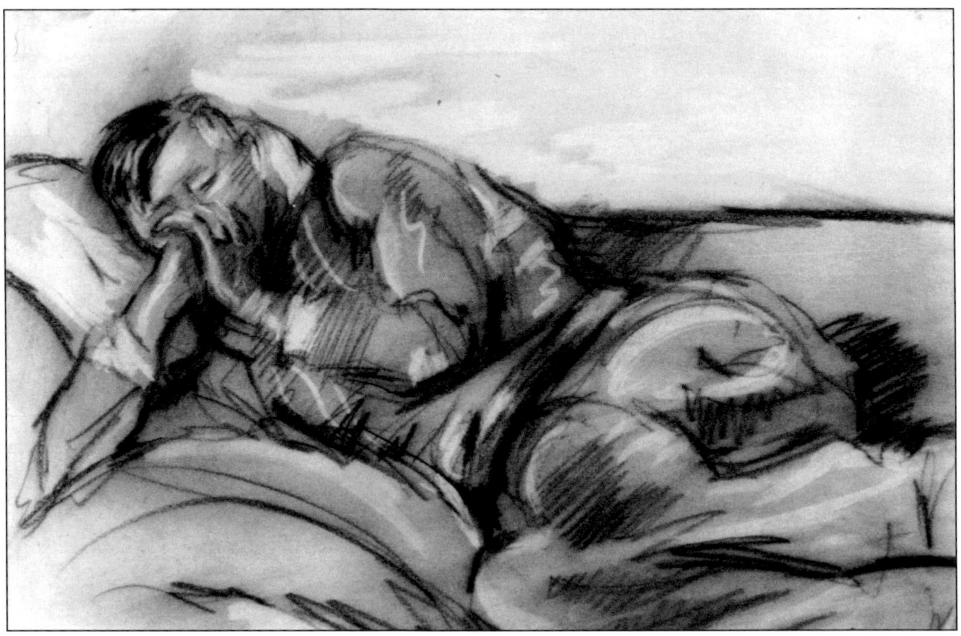

Sleeping Miner - Charcoal on Paper 1947

Tom's work began to be widely collected and shown, both in group and individual exhibitions, throughout this country and abroad. The 1970s saw a significant change in the organisation and promotion of his work commercially. In 1968 Lady Hirshfield had purchased some of Tom's paintings whilst she and her husband, a Labour peer in the first Wilson government, visited the Durham Miners' Gala. They were so enthusiastic about his work that Desmond Hirshfield proposed a contract of co-operation to Tom, valid for three years. Under the terms of this agreement, Tom was to receive 45% of the net sale price. Lord Hirshfield, as sole agent, would meet the exhibition gallery's commission, all expenses incurred in arranging the exhibitions and sales, including framing costs, together with the cost of printing and promotion of the exhibition from his 55% of net receipts. This arrangement was to have the effect of successfully promoting Tom's work nationally and soon exhibitions in London were to follow. It did, however, have the detrimental effect of forbidding Tom from exhibiting or selling work independently.

Tom was delighted at the prospect of exhibiting in London. "It is what every artist dreams of ... my painting has been recognised in Northern art circles for a long time. Now the London show will give me a national outlet, and maybe put me on the international scene. But I'll go on working in mining as long as there is a pit here. Art is a bit too uncertain to make a living from," said Tom in 'Coal News' January 1972. Despite his success and the media interest Tom still saw himself as a workman, a family provider and above all a miner who could paint rather than an artist who was trapped in the stony grasp of the coalmine. By the time the John Whibley Gallery exhibition opened in Cork Street, Mayfair, only two of Tom's thirty seven paintings remained unsold.

Lord Hirshfield, who had originally encouraged Tom "to paint what he liked and whatever he wanted to so as not to become commercialised," by contrast, in November of 1972, told Tom that "Private buyers of art in the south of England prefer your above ground paintings (views of collieries, street scenes, landscapes etc) to the underground paintings of miners working in the pits ... the subjects are clearly a little strange to those who are not very conversant with the mining industry and miners' working conditions." He went on to suggest that smaller paintings in brighter colours were more commercially acceptable to the London market. In response to Lord Hirshfield's suggestion that he 'broaden his palette to include other industrial subjects and beach scenes' Tom responded, "I tried to illustrate ship building at Wallsend but you have to live by the river to know it inside out." Almost thirty years later, he was reaffirming Bill Farrell's advice of 'paint what you know' - Tom needed to remain true to his original artistic principles.

Tom resisted what must, at the time, have been a very powerful external attempt to influence the direction of his work. Jack Reading, a collector of art and sponsor of mining art observed 'It wasn't sincere and so he went back to what he knew, and he was right.' Jack Reading's comments were earnestly and kindly meant. When Tom had painted beach scenes as suggested by Lord Hirshfield he used his wash-off technique, a technique which was not as suited to the depiction of the sunning bather as it was to the representation of the blackened miner. Indeed, ironically, many of the resultant beach scenes (Plate 16) were dominated by the complex geometrical structures of the piers of Redcar and Saltburn or of Blackpool Tower, recalling shapes so familiar to Tom and mirroring the structure of the pithead gear - perhaps Tom had not moved as far away from his central subject matter as Lord Hirshfield had intended.

The agreement was ultimately terminated in October of 1974 when Lord Hirshfield informed Tom of the then very serious and adverse business climate. Whilst their business adventure had had a positive influence on Tom's artistic profile, he had recognised its unacceptable influence on his subject matter. Tom's profile was further raised when The National Coal Board featured him in two issues of their news film 'Mining Review' and his work was also the subject of a BBC 'Omnibus' documentary. This resulted in a prestigious commission from Barclay's Bank. Tom was always reluctant to accept commissions, preferring to work inspirationally, but in 1976 he was happy to undertake this commission on behalf of the bank, sponsors of Durham Big Meeting. This large oil painting entitled 'The Miners' Gala' (Plate 19) hangs in their Durham Market Place branch to this day.

The severing of the business arrangement between himself and Lord Hirshfield was not the end of Tom's London adventure. In 1978 he organised his own exhibition at Brampton's Gallery, in Marchmont Street, Bloomsbury, crating up seventy seven of his works and despatching them to London. He had seen the gallery's advertisement in an arts magazine and determined to mount another show of work in the capital. This was to be the first time any of his lithographs were exhibited. This exhibition was equally as successful as his Whibley Gallery exhibition and gives testimony to Tom's drive and self belief. It was to have far reaching consequences. The exhibition was viewed by a reporter from the Soviet news agency Tass. As a result Russian newspapers were very complimentary about his work. This Russian connection was to culminate in an exhibition of works by Northern Artists in Moscow in the early 1990s which included work by Tom.

The Development of the Characteristic McGuinness Style

Tom had a 'love-hate relationship with the colliery'. He was phlegmatic enough to realise that the mining industry provided him with a good wage and provided his family with a comfortable and guaranteed living - a living which he felt his art could not underpin. He also had a great regard both for the dangers of the coalmine and for the constant support of his workmates. This emotive ambivalence of fear, respect and awe is reflected in the empathy and the mastery of his art.

It was during these early years that Tom's characteristic style began to evolve. His life drawings demonstrate his tremendous skill as a draughtsman. Gradually he introduced the artistic device of distortion to create a more powerful and emotive effect. Distortion of all four elements of the picture - line, tone, colour and composition - has created the easily recognisable McGuinness style. This is done not to caricature or lampoon his working colleagues but to give expression to the depth of feeling the artist experiences when depicting the miner's lot. This emotional portrayal defines the essence of expressionist art. "McGuinness has evolved a style which is linear, intensely rhythmic and very luminous in colour, due chiefly to the use of glazes. The juxtaposition of distortion, rhythm and intense luminous colour produces a feeling of surrealistic nightmare. McGuinness has an important attribute, almost theatrical in concept - a regard for gesture. His miners yawn, chew and spit ... his paintings make their point - one is never in doubt as to his feelings about the subject of his paintings - a true Expressionist," observed Scott Dobson in an article in the Newcastle Journal of 1966.

Tom works in a variety of media. Originally using charcoal, conté and pastels, he began to paint in oils in the 1950s and his masterly use of glazes, which he developed in the 1960s, gives his underground scenes their characteristic sepulchral, ethereal quality. In his emotive glazed oil 'Men in the Cage', of 1961, (Plate 12) Tom powerfully demonstrates the impact of these techniques by carefully selecting tones from a limited spectrum of colour to convey the utter exhaustion of the ascending miners. The luminosity of the blue glazes reflects the atmospheric, chilling damp which envelops the huddled figures as, cowed and perished with cold, they turn to each other for mutual solace, anticipating release from the confines of the mine and the entrapment of the cage. "... he developed a way of conveying the coldness and shine from the coal," said Nerys Johnson, Curator of Durham Light Infantry Museum and Art Gallery.

In his constant search for new ideas, styles and techniques Tom was drawn towards experimentation with printmaking. He had always studied art literature, a legacy from his Settlement years, and in the early 70s began to refine his self-taught etching style, particularly following the classical principles and methods of Albrecht Durer. It is difficult to know how much Durer's work influenced Tom but he did know of the Flemish Renaissance artist's book on human proportions and of the spiritual and bellicose symbolism in Durer's work. The precision and formality of line in Durer's etchings appealed to Tom and led to his appreciation of the more random incisions in Rembrandt's etchings - himself a great admirer of Durer. This was to lead Tom to study the powerful etching style of Goya - an artist with a bold, almost random approach to etching, but an artist whose images conveyed the darkest and most disturbing meaning. Whilst it is the dark subject matter of many of these works which intrigued Tom, Goya's use of aquatint, giving tonal qualities to his etchings, particularly excited him. Tom follows precise and almost classical traditions in his etching technique, adapting and personalising the process to produce prints on his own cast-iron press.

Becoming interested in lithography, in 1976, he was awarded a Northern Arts Print Bursary which enabled him to hone his lithographic skills at Lowick in Cumbria under the tutelage of John Sutcliffe. Tom's natural free-flowing style lends itself to the immediacy inherent in the lithographic process where the image is drawn in a waxy medium onto the printing surface. During the ensuing years he has developed this technique to create some of his strongest and most evocative images. The flowing lines of Tom's lithography can be seen in his 1977 image 'Harvey Men Going Outbye' (Plate 21) in which Tom places the observer in a uniquely low, eccentric vantage point. This dictates that all aspects of the picture's perspective curve away from the viewer's focal point. The miners themselves react to this artistic distortion with their gyratory progression from their day's work.

Much of his recent work involves mixed-media - ink, watercolour and gouache - Tom's 'wash-off' technique. He has adapted this method and uses it to splendid effect to give density and depth to the scene. Whilst less precise than oils the effect is equally dramatic. The apogee of this method is 'Resting on the Way', his 2004 study for a stained glass window in St. Joseph's RCVA School, Coundon. Here Tom movingly represents human experience as he places Mary and Joseph, in modern dress, resting on a park bench at the edge of a mining village. Joseph is shown as a miner complete with his Northern Echo and bottle of Brown Ale. Mary, attending to the infant Jesus, is resplendent in headsquare and mackintosh, whilst a pit pony, replacing the donkey, takes a well-earned rest. The Holy Family in the foreground contrast with the distant features of the colliery, the school and the church with its graveyard. All are cloaked in an inky-black sky from whence the trade-mark McGuinness moon bestows benign guardianship over the scene. The miraculous emergence of the residuum of colour creating a translucency following the wash-off procedure is in remarkable contrast to the intensity of the indian ink used - giving a sense of layering and depth to the picture.

Although in recent years the scope of his work has broadened to include other themes, including religion and a series of semi-abstracts painted whilst artist-in-residence at the nuclear power station in Hartlepool, he continues to document the decline of the once-thriving mining communities. Throughout his career Tom has remained constant to the original source of his inspiration - the coalmine. Despite having been made redundant from mining in 1983 he still paints daily, gaining inspiration from his drawings, his memory, his imagination, and from his copious sketch-books.

The eighties, however, was a decade of great trauma and great creative change for Tom. Sadly, in 1982 his wife Cathy died after a short illness. Tom confided to his friend Roy Thomas that with Cathy gone he had lost his inspiration. Despite it being a very unusual pastime for a working man, she had a strong belief in his artistic ability and an empathy with his need to paint. Tom had married Catherine Quinn, eleven years his junior, in 1959. He confirmed that Cathy continually encouraged him to draw. She was always very supportive, even when times were hard and he required money to buy materials. He was even allowed to use the spare bedroom as a studio and she would keep the children occupied to allow him to draw in peace. Many family studies survive from their early married life - Cathy nursing the children, sewing or working at domestic tasks. Tom derived huge satisfaction from capturing the personal moments of his own close family life.

Tom remains unsure whether or not the sad and untimely death of Cathy influenced his art, but his daughter Corinne observed that her father's work became more introspective at the time although his output continued, possibly due to the encouragement of Roy Thomas who convinced Tom that it would have been Cathy's wish. Tom's elder son, Shaun, confirms that religion had been very important to his mother. She was a great supporter of the local church, St Mary's, Bishop Auckland, whereas he concedes that for Tom church attendance was a time for reflection. However, it was at this time, shortly after Cathy's death, that Tom accepted a commission by the priest in charge, Father Michael Finnigan, to design a series of stained glass windows and the tabernacle for St Mary's. He would not accept a fee for this commission. (Plate 31)

The Catholic religion has been a huge influence in Tom's life. He was brought up in a powerful Christian tradition in a village where religious identity was very strong. Because of the cosmopolitan make up of Witton Park's inhabitants due to the sudden influx of workers from all parts of the United Kingdom and abroad, many faiths prospered as each group imported its own brand of Christianity. Thus a village with four thousand souls could boast eleven different permanent churches and chapels as well as a travelling chapel. The sense of belonging, if not segregation, was all-consuming.

St Chad's Roman Catholic School was not merely the focus for Tom's primary education it was also the location for his spiritual instruction as the school was for many years used as Witton Park's Catholic Church, serving as such until the opening of St Mary's Church in Bishop Auckland in 1956. Father Walmsley, Witton Park's local priest in Tom's early years, was an important figure in his young life. He is remembered as a kindly man with a sense of fairness to match his sense of humour. Father Walmsley gave Tom his first commissions - to portray popular racehorses of the day. More significantly, here was a respected authority figure taking a personal interest in Tom's welfare and in his talent.

Tom was thus extremely proud and honoured to receive the St Mary's commission, the designs for which he did very much in his own style. He deliberately introduced contemporary images into the biblical scenes, notably in his third window of the series, 'The Last Supper', in which Christ's head is 'adorned' by a 'halo' of colliery winding gear. In the original designs for the panels for the tabernacle the scene in which he depicted the betrayal in the Garden of Gethsemene featured riot police - an implicit reference to the police action in the 1984 Miners' Strike. Tom, who had been made redundant from mining in 1983 after thirty nine years of service, was about to make a political statement as his world had totally altered in the space of one year. He was later to change his mind and deleted them from the finished design.

Art historian, Rosemary Davidson, commenting on Tom's religious works, admired his placing of biblical events in a contemporary setting which she linked to an artistic tradition which could be traced back to the Renaissance. Rosemary affirms that he is part of this long tradition which has continued into the 20th century and she exemplifies this by comparing Tom's work with that of Sir Stanley Spencer and his iconic image 'The Resurrection in Cookham Churchyard'. Interestingly, Tom continued to produce religious images throughout the 1980s which echoed this anachronistic iconology - the juxtaposing of the contemporary in an historical setting. In his mystical 'Crucifixion' series of etchings, 'Number 2' shows the crucified Christ against a backdrop of the skyline of a terraced pit village with its winding gear and slag heaps. The flat-capped miners and their head-scarved wives stand in silent homage.

At about the same time, Tom was sponsored by the Arts Council as Artist in Residence at Hartlepool Nuclear Power Station. His brief, alongside one hundred Cleveland art students, was to use the power station as a 'stimulus and inspiration'. The resultant twenty images produced during his tenure were very different from any of his previous works. Admittedly, the subject matter was different from what had gone before but, given this difference, why did he radically change stylistically? There seems to be a significant move towards abstraction in the body of work produced - one in which reality is subsumed by design. Had he really 'lost his way' - as he had confided in Roy Thomas? Was he seeking new directions following the death of Cathy? Was he disillusioned with the political determination to destroy the coal industry or did he indeed derive inspiration from his different surroundings? Tom, not surprisingly, is unable to determine the reasons for his stylistic change but out of a time of great turmoil came a fascinating move towards the spiritual and the ethereal.

In his Power Station Series (Plate 33), Tom contrives, from a limited palette, to explore his vision of the ultimate in man's generation of power. The energy derived from coal is locked underground in an enclosed, claustrophobic environment - an environment that Tom knows and has portrayed in his inimitable, expressionist style. In contrast, energy derived from the intangibility of nuclear power, above ground, in the light, to Tom, cannot be portrayed using the same McGuinness trademark of vibrant colour. Tom's solution was to depict the potential of this power-house of energy in an abstract manner, in complete contrast to the realistic, representational style of his portrayal of the religious images he used for his stained glass windows at around the same period. The radically different approach to the Power Station Series can also be appreciated when compared with the coloured lithograph 'Machine Men Starting the Cut' (Plate 28). In this image, Tom returns to familar territory, the underground world. The bold depiction of the sophisticated self-advancing cutting equipment is confidently displayed in all its detail giving an impression of great depth to the coal face - the sense of design is in the subject not in the interpretation.

These two new bodies of work, his religious images and the Power Station Series, were shown in complete contrast to each other in an exhibition at the Dovecote Arts Centre in Stockton in 1985 under the title 'The Power and the Glory'. One can speculate that had Tom not been made redundant and had he not had a close personal bereavement in the early 1980s neither of these important bodies of work would ever have been produced. They are so radically different from any of his previous or subsequent work that one can assume that these traumas did influence his artistic direction. By the end of the decade Tom had reverted to his familiar theme of the coalmine, using his sketch-books and personal experiences to produce his 'Lost Generation' series.

As the North East mining scene changed in the 1990s, so too did Tom's work - as reflected in The Lost Generation series - a series in which he traces the physical destruction of the industry, together with its social consequences. From the determined demolition squads and salvage workers through to the now redundant miners and their families Tom affirms that the physical destruction of the colliery has destroyed a way of life. The miners are still bowed, not with toil, but with sadness. The power and influence that the mine exerted still lives on. "They've closed all the pits down and people are more and more living on memory," says Tom. "The Lost Generation is a series about the aftermath of what's happened. It shows groups of miners wandering around aimlessly and going back to the pit heap. At one time everything was left and nature took over, covering the place with weeds. That was good - it created atmosphere."

The atmosphere Tom deliberately creates in this series is one of revulsion. Rarely do we see faces and facial expression in Tom's work but here we are haunted by images of survival - emaciated, etiolated men whose hollow eyes glare out at us like survivors from a concentration camp, silently pleading that we bear witness to the consequences of their struggle to win our coal. Their accusatory stare demands that we consider the irony of their plight - the deliberate destruction of their industry, an industry which inflicted so much physical degradation and morbidity, yet upon which they depended for their very survival.

This is exemplified in 'Old Men on the Heap,' (Plate 40) a 1996 oil in which the victims of redundancy plead their case in front of the slag heap. Their three shaven heads are mirrored in the three peaks of the dark waste heap which is their ultimate backdrop. Whilst the sky appears almost aflame, the redundant headgear looks impassively down on their skeletal faces, no longer guardian of their welfare. Tom succeeds in making the viewer feel as uncomfortable as do the three gaunt faces that stare out from the canvas.

This imagery is taken to extremes in Tom's 1986 etching 'The Decline' (Plate 34) which depicts a spectral skull hovering over a redundant colliery whose leaning pit frame and pulley wheels seem to be melting into the surrounding night. When this image was shown in the 1992 Woodhorn Church exhibition at Ashington it provoked Stephanie Brown in Visual Arts Review to comment "least comforting of all (the images) is a small etching of The Decline ... Made in 1986 this print provides a chillingly prophetic glimpse into the threats lying over the pits today ... Tom's work is overwhelmingly concerned with harsh realities ... the cramped, tomb-like spaces, vertiginous heights and plummeting depths. These are observations of an insider, one whose ambitious technical and conceptual range extends the work beyond the particular and parochial into more universal thought-provoking territory. There are no cosy reminiscences in McGuinness's painting of a pit village sandwiched between a cemetery and a slag heap, which looms above it like some bleak Golgotha, or in the cadaverous faces of miners going home."

Even though Tom's miners realise they must eventually move on, the lure of the industry continues to haunt them. They are unable to break free. Ironically, the lure of the industry still haunts Tom. Although he had the opportunity to break free in the 1980s, he also found the lure of the mines too compelling. It drew him back.

By the mid 1990s Tom had absolved himself of his obsession with the plight of the Lost Generation. This manifested itself in a lightening of his artistic mood and the strengthening of the colours of his palette. It was a prolific period for him. He produced many large colourful, vibrant oils in which he deliberately extended his colour range with the notable use of cadmium yellow, vermilion and ultramarine. In Harvey Wetmen (Plate 44) - a large glazed oil painted in 1996 - the figures again turn away from the observer, they no longer need to seek the support that eye contact with the viewer confers, as their glistening vivid yellow waterproof jackets contrast with the bright moonlit sky. The hunched figures shuffle their way past the gaunt headgear and angular conveyor towards the welcoming lights of the lamp room. Tom's trademark moon again assumes guardianship of the night scene. Tom is back on familiar ground and his work for us, the observer, and for the Harvey Wetmen seems more reassuring.

Art historian Elizabeth Conran contends that there is a lyricism and a spiritualism to much of Tom's work. Tom's beliefs and commitment to Catholicism could be evidenced as supporting this contention. If we analyse Tom's imagery from this point of view then his choice of composition and symbolism become more apparent and the iconology of his work is revealed. Much of this may be, for Tom, at a subconscious level and thus intuitive. This can be exemplified from several of Tom's better known paintings. 'Early Morning Shift' (Plate 10), owned by Middlesbrough Art Gallery, is one of Tom's personal favourites. He explained that the image was conceived when he was riding in the early morning train to Darlington, crossing the embankment above Eldon, looking down on the village and the Black Boy Colliery. The impact of the scene was such that it firmly implanted itself in Tom's mind and he later committed it to canvas.

Tom was not faithful to the scene. He depicted the colliery with a third winding headgear and with three chimneys. This may have been mere artistic licence but equally may demonstrate deliberate symbolism. These additions to the lowering, foreboding backdrop of the colliery slag heap have resonance with Calvary and the crucifixion. The miners going to work become self-sacrificial martyrs. The mine promises economic salvation for the many but represents destruction for the few in the same way that the Christian belief interprets the consequences of the crucifixion as salvation for the many but with a resultant selfless sacrifice for the few. The icing sugar roofs, which may simply represent a frosty night, metaphorically remind us of the fragility of human existence and the elevated viewpoint confirms the supernatural iconology of the image. We know that the viewpoint is from the railway embankment but Tom gives us no reference point in the painting to make this explicit. Is this a deliberate device or does Tom subconsciously and intuitively see a religious analogy, otherwise what was it that so impacted upon the sleepy artist in that misted morning train?

In the same way, we can consider the iconology and the composition of Tom's 'Miner and Child' (Plate 1) - now in the Shipley Art Gallery, Gateshead. It is widely agreed that this image can withstand the scrutiny of artistic criticism purely on merit. Have we ever seen concern, that most subtle of all human emotions, portrayed so tenderly as in the expression of Arthur Raine as he soothes his ill child who responds by tugging on his collar? But what was it that drew Tom to take up crayon and paper to capture this moment of domestic trauma rather than offering to help Arthur in a more practical way? Could it be the subconscious symbolic religious imagery which caught Tom's artistic and interpretive eye?

The drawing's structure is familiar to us, even if the characters are not. It withstands direct comparison with the traditional portrayals of the Madonna and Child in its composition and the revealed level of concern. There is a striking similarity of pose and emotion. This is a classical composition which has been used by generations of artists to symbolise all-encompassing and unconditional love. Surely it is this religious connotation that so appealed to Tom in such a tender but worrying moment?

The religious connotation is also apparent in his 1992 coloured lithograph 'Women Waiting for News' (Plate 37). Four young women stand expectantly outside the colliery gates waiting anxiously for news of their men's plight following a pit accident. The group, placed firmly against the background of the colliery winding gear, are being approached by two ghostly figures bringing the news. Resonances of the crucifixion scene, which places the Marys waiting at the foot of the cross for news of the crucified Christ and his promised resurrection, are apparent in the anguished faces of the four women and the insubstantial shapes of the would-be messengers. Have their men survived the entombment of the presumed colliery disaster? Will they be brought back alive to them? Tom places four women at the colliery gates - was he aware of the legend which places a fourth Mary, Mary-Jacoby, sister of the Virgin, with Mary-Salome, mother of the apostles James and John and Mary Magdalene to share the grief for the crucified Christ? In conversation Tom confesses he was not aware of the story but the imagery remains equally valid. Such is the power of the scene that Tom continuously revisits this theme of women waiting for news.

Art historian Rosemary Davidson suggests, "We tend to think of the mining life as a male preserve but in Tom McGuinness's work the women of the mining community play a significant part. He sees them always in small groups, sometimes with children, never with men; always in relation to the road to the pit, to the pit head itself, to the gates or to the yard. Tragedy looms for the women who wait for news in works that recall Kathe Kollwitz in their Expressionist power ... in more recent pictures, desolation descends on the women of the lost generation who have seen the context of their life destroyed."

By no means is all the iconology, the hidden meaning, in Tom's work religious, he makes political and economic statements through the medium of his art. In his 1990 etched and aquatint print 'The A Team' (Plate 36), a title coined by Tom to represent a group of 'top workers', we see the men discussing which tunnel to enter - the old tunnel with its wooden props and labour-intensive working methods or the tunnel on the right with its highly-mechanised power-loading equipment. A dilemma indeed - jobs or efficiency, short-term pragmatism or long-term security? However, if we look beyond the group we can see the darker figures of the deputies in the background. They will make the decision that will dictate the future of these 'top men'. Tom is making a deliberate economic statement, through the symbolism in the image, about the future of the industry and its workforce. He is also making a political point - the future is in the hands of the decision makers - the politicians. The industry was at a crossroads. This is a prophetic and profound statement for, in 1990, the last of the great rounds of pit closures in County Durham was just beginning. By 1993 the last coalmine in the county had been closed down. Economics and efficiency had triumphed over loyalty and humanity.

Chris Lloyd of the Northern Echo confirms, "They are now erasing the mines from the landscape. On Fishburn where Tom started there is a golf course: from pithead baths in Gateshead there will soon be springing an angel sculpture but they can't erase the mines from Tom's memory because Tom's works remain to recall life both above ground in the tight pit communities and below ground in the tight pit tunnels."

Political and religious interpretations are by no means the only observations that can be inferred from the work of Tom McGuinness. As an expressionist artist Tom heightens the emotive power of the image. In Tom's chalk, wash and ink drawing, 'Filler on the Face' (Plate 8) - executed in 1958 - a solitary miner, is gainfully employed vigorously filling tubs from the loose coal that has been blasted from the face. Here the collier is rounded, robust almost heroic. He apparently feels secure and Tom's use of warm colours makes the observer feel as comfortable with the image as he obviously feels with his task. Whilst located within tight confines, his form has moulded to match the shape and dimensions of the tunnel in which he operates. Man, the machine, becomes master of his environment.

This image is in direct contrast with 'Miner on the Face' (Plate 13), a large painting executed in the 1960s in egg tempera, which demonstrates Tom's concern for safety and indeed his fear of the dangers. It may reflect his changing attitude to his working environment. Here the solitary miner, apparently abandoned like the old workings in which he finds himself, is compressed into a kneeling position and appears to be trying to support the sagging roof with his left cheek. He is unable to hold the weight of the lowering roof as his elbows buckle, mirroring the bending, breaking wooden props that surround him. Affording even greater power to the nightmarish scene is his grey skin, blanched by the adrenalin of fear. His left hand spasms into a talon as he tears his finger nails through the hard earth in an attempt to claw his way to freedom. The medium chosen, egg tempera, adds to the stark clarity of the image, as its unyielding nature does not easily allow blending and grading of the colours, reinforcing the trauma of the miner's plight. The vigorous and controlled movements of 'The Filler on the Face' has been subsumed by the blanched, angular, ashen skeletal form of the lost and frightened 'Miner on the Face'. Man has become vulnerable; mastered by his environment.

Tom takes this imagery even further in his ghostly almost mystical 1996 oil 'The Coal Hewer' (Plate 42). The effect imposed on the viewer is that of an ethereal and spiritual being, a miner from the past - a tragic symbol of a lost industry - petrified by his plight. Tom has given this hand-hewer a spectral quality as he picks at the canch - the loose overhead stone - one of the most dangerous aspects of working on the face. The hewer, seeking refuge in the left half of the picture, nevertheless is white with fear. Tom, by his inimitable use of glazes, introduces a blue-green lustre to the scene, heightening the tension and involving the observer in the drama. In the half-light a look of horror is etched into the hewer's face. His anxiety is contagious, infecting the observer with dread. He manifestly does not want to be there. Despite our revulsion we are compelled to be present at his demise - unwilling witnesses to his fate. Man has become victim of his environment.

Epilogue

Tom has continued to exhibit to wide public and media acclaim. This media interest culminated in 1997 when Tom's work received enormous public approval in a major retrospective exhibition, 'Mines a McGuinness', shown at the Bowes Museum at Barnard Castle. This coincided with the publication of the definitive book on his life and work 'Tom McGuinness - The Art of an Underground Miner' - McManners and Wales. The exhibition toured both nationally and internationally. Further successful exhibitions followed - 'McGuinness is Good for You' (1998), 'The Power and the Glory' (2000), 'The Shaft' (2002), 'Face to Face' (2004) and 'Etched in Coal' (2005) - the first exhibition to feature all of Tom's printed works. In addition, over fifty examples of Tom's work - the Jack Reading Bequest - are permanently held at Bishop Auckland Town Hall where they are exhibited annually in the gallery which now bears his name. Tom continues to show and sell his current work in solo exhibitions throughout Great Britain and beyond. His association with Richard Hindle of Gallerina, a commercial gallery in Duke Street, Darlington, has led to a series of highly successful selling exhibitions which generate huge public interest. To date he has exhibited in over fifty solo and over forty joint exhibitions at home and abroad. His work is represented in galleries and private collections nationally and internationally.

From 2005 Tom had the opportunity to rekindle his love of printmaking with the installation of his printing press in The McGuinness Gallery at Bishop Auckland Town Hall. He is keen to share his vast knowledge of this artform and regularly holds informal workshops for all abilities and all ages. In addition he has experimented with computer technology, regularly creating original digital images. Thus his continual quest for knowledge and self fulfilment even extends to mastering the most modern technology.

His profile has been raised on the national art scene with the inclusion of his entry in the definitive 'The Artists of Northumbria' by Marshall Hall in which Tom's iconic drawing, 'Miner and Child', is prominently displayed on the frontispiece of the book. A self-portrait featured in the national touring exhibition 'Face to Face: British self-portraits in the twentieth century'. The collection consisted of one hundred portraits of artists that had been collected by Ruth Borchard in the late 1950s. Ruth approached Tom in 1958 for an image of himself, paying 15 guineas for the painting. Tom also featured in the accompanying book that was edited by Philip Vann and featured such notable artists as Lucien Freud, Graham Sutherland and Francis Bacon. The biographical notes state, "Here is a man in whom a strong sense of identity - conveyed partly through the dynamic black outlines throughout - co-exists with acute perceptiveness."

As a painter of the industrial scene Tom is unique in that he continued his profession as a miner throughout the whole of his working life. He sees the world in a distortion of reality. His work colleagues are bent, almost cowed, by their awesome task. Their twilight, subterranean world is illuminated by an eerie, ethereal light which adds a spiritual dimension, foretelling a prophetic transience to an industry that at one time seemed indestructable.

"Great writers and artists of the past have shown us that much beauty and meaning can be revealed in the tragic and so it is evident in Tom's work. Because he can raise his paintings far above the mere representational, far above the mere accurate record of coalmining, they become valuable works of art. I believe Tom has much in common with many major artists of the past," asserts Gill Holloway.

"I find it difficult to express myself in words that is why I paint. I put my feelings into my paintings and my art mirrors my life in the mining community." Tom McGuinness

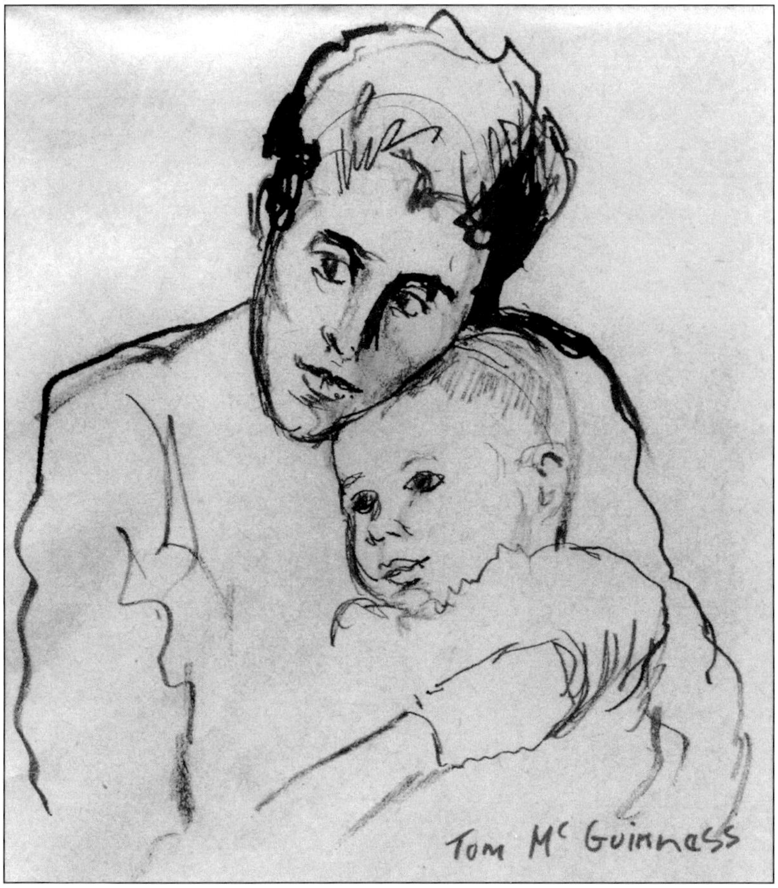

Cathy and Shaun - Ink Sketch 1960

The Plates

Plate 1

Miner and Child 1948

Conté on paper - 45 x 33cm

Courtesy of The Shipley Art Gallery, Gateshead

Purchased by the Shipley Art Gallery, Gateshead in 1949 from the Federation of Northern Arts Society exhibition held in the Shipley Gallery in March of that year. G. N. Drinkwater, the gallery curator, told members of the committee that it would hold its own in any gallery in the world. The purchase price was ten guineas.

See also pages 13 & 24

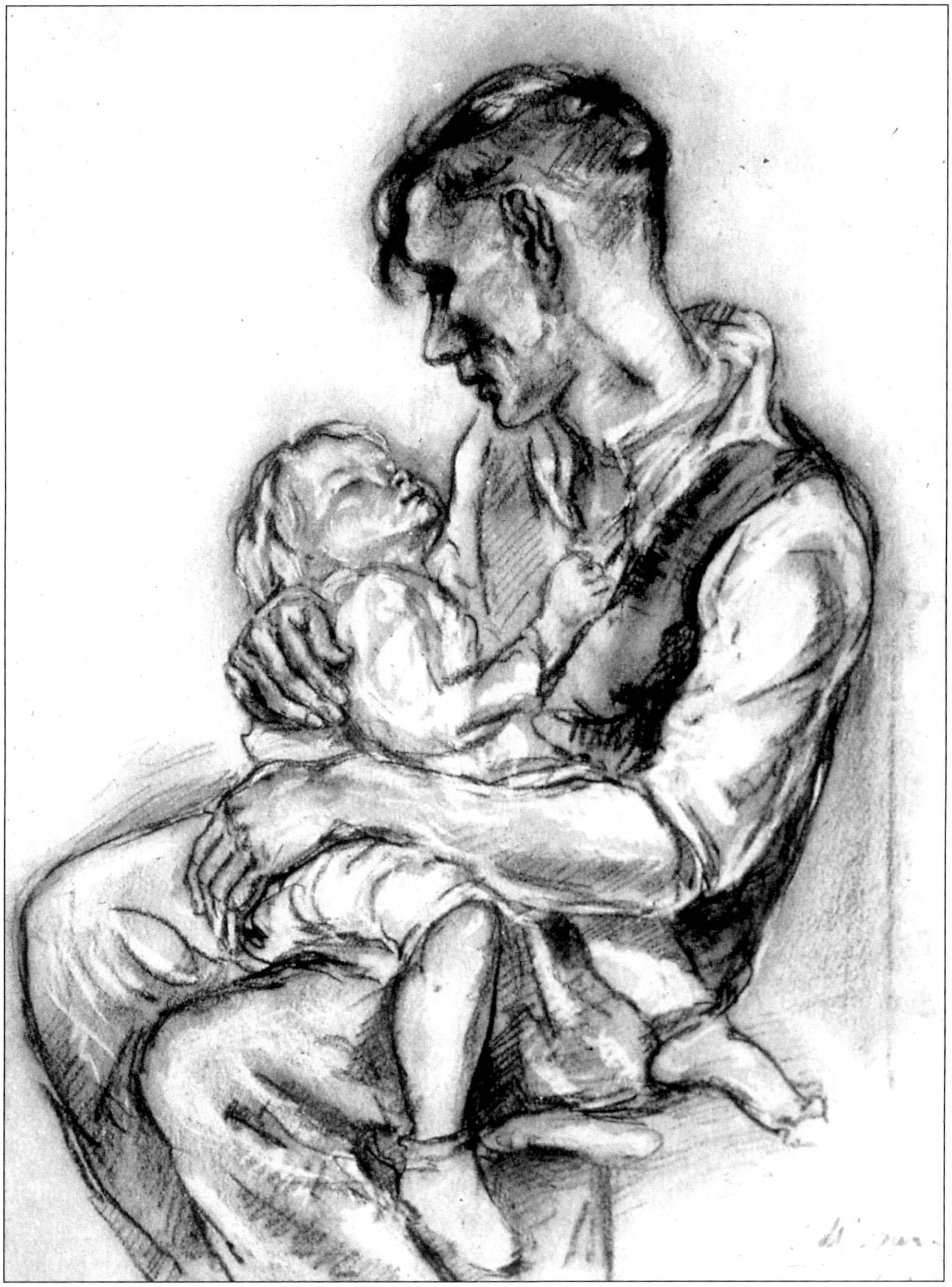

Plate 2

Miner Undressing 1948

Conté on paper - 40 x 28cm

The subject once again is Arthur Raine who lived opposite Tom in Thompson Street, Witton Park. It was Arthur who was instrumental in getting Tom a job in Wooley Colliery, Stanley, Crook. Now part of the Jack Reading Bequest, this picture was exhibited in the national touring exhibition 'Coal' in 1983.

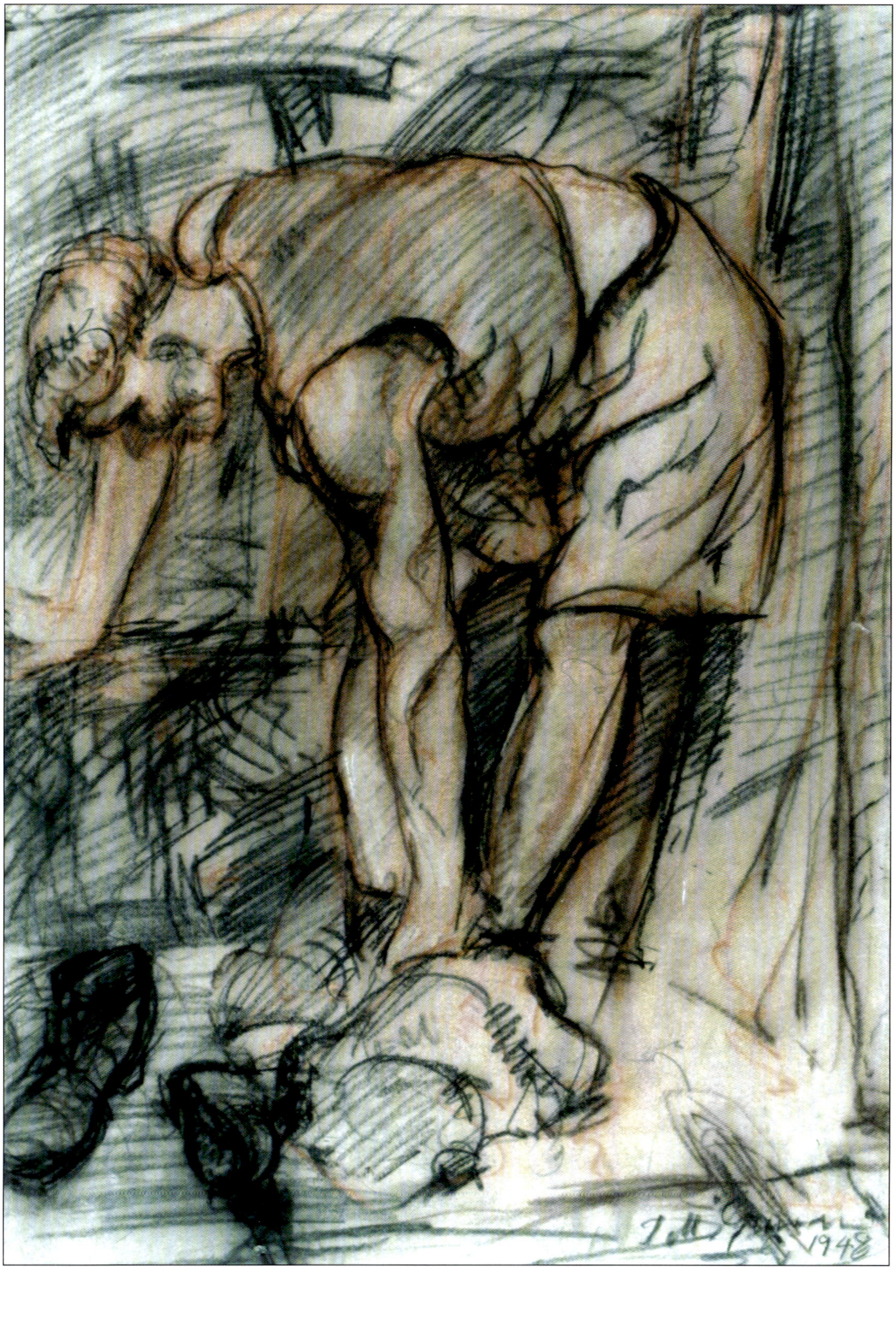

Plate 3

Life Study 1950

Pastel on sugar paper - 55 x 38cm

Enthusiastic art teacher Ralph Swinden, based at Darlington School of Art, wanted his students to be proficient in anatomical drawing. Being unable to afford models they made life drawings of volunteers from the class. Tom gave this drawing to his fellow artist Tom Alderson during their time together at The Spennymoor Settlement.

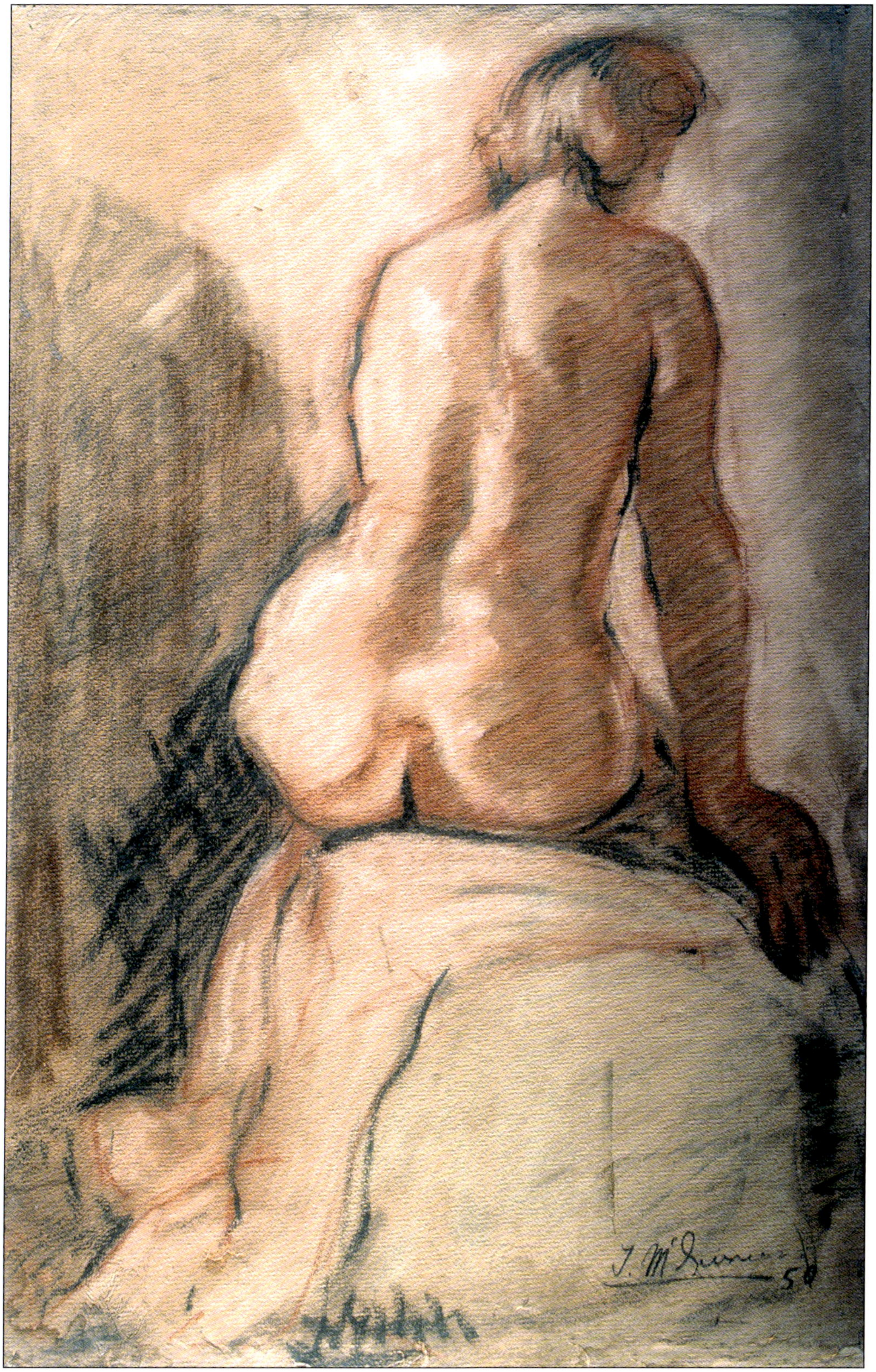
T. McGuinness 50

Plate 4

Hand Putter 1951

Unglazed oil on canvas - 40 x 60cm

One of Tom's earliest oil paintings and one of his favourites, Hand Putter always hung in pride of place above the fireplace in his living room. Heeding Spennymoor Settlement Warden Bill Farrell's advice to 'paint what you know', Tom used his personal experience of putting from his Bevin Boy days. The figure, almost contorted to the shape of a question mark by the confining canvas, is one of the earliest examples of Tom's use of distortion to heighten the experience for the observer - one of the elements of expressionism.

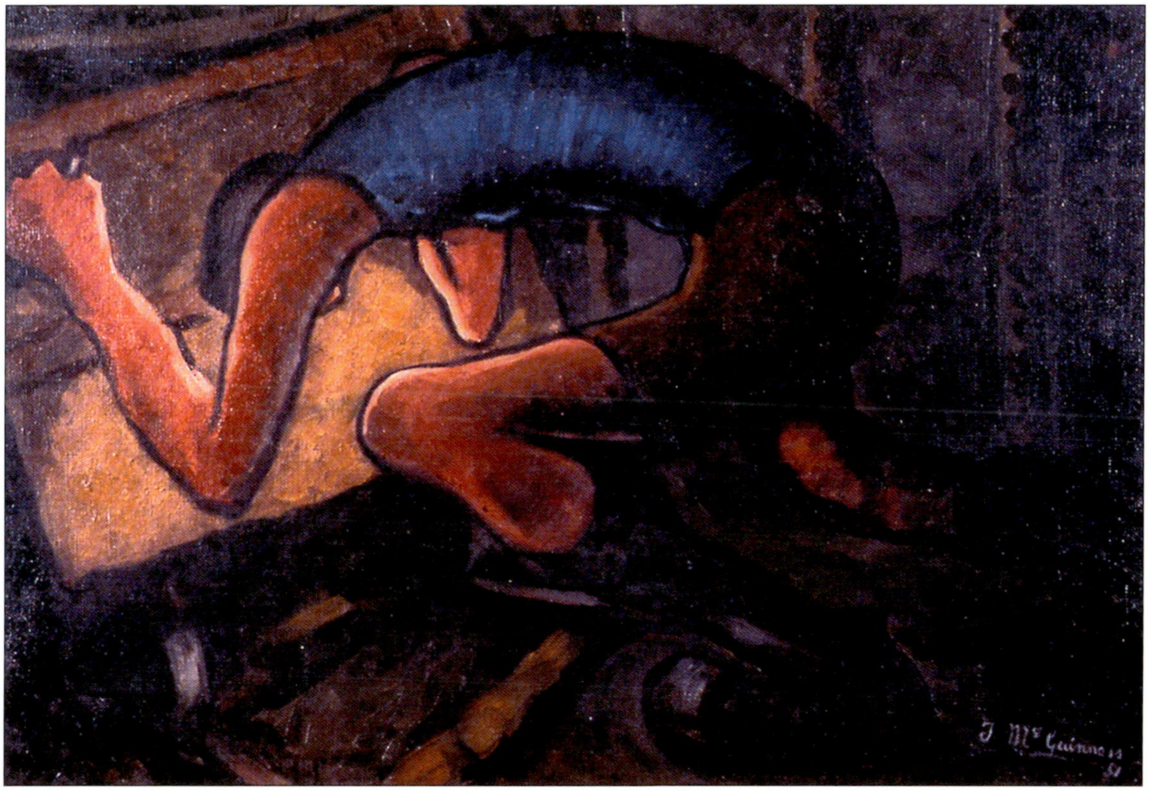
T McGuinness

Plate 5

Pigeon Crees 1953

Oil on canvas - 46 x 63cm

Accepted by the Royal Academy but never hung, this painting was originally presented to the National Coal Board Sports and Social Club, Hobart House London. Showing the influence of his time at The Spennymoor Settlement, this realistic alla prima study depicts miner Havelock Hope, later to die from lung disease in his forties, waiting for Jack Tray's homing pigeon. The Witton Park Salvation Army Citadel is on the right. This image was used in David Crouch's book, 'The Art of Allotments: culture and cultivation' and as a poster to promote the musical 'The Allotmenteers'. Pigeon Crees is part of the Jack Reading Bequest.

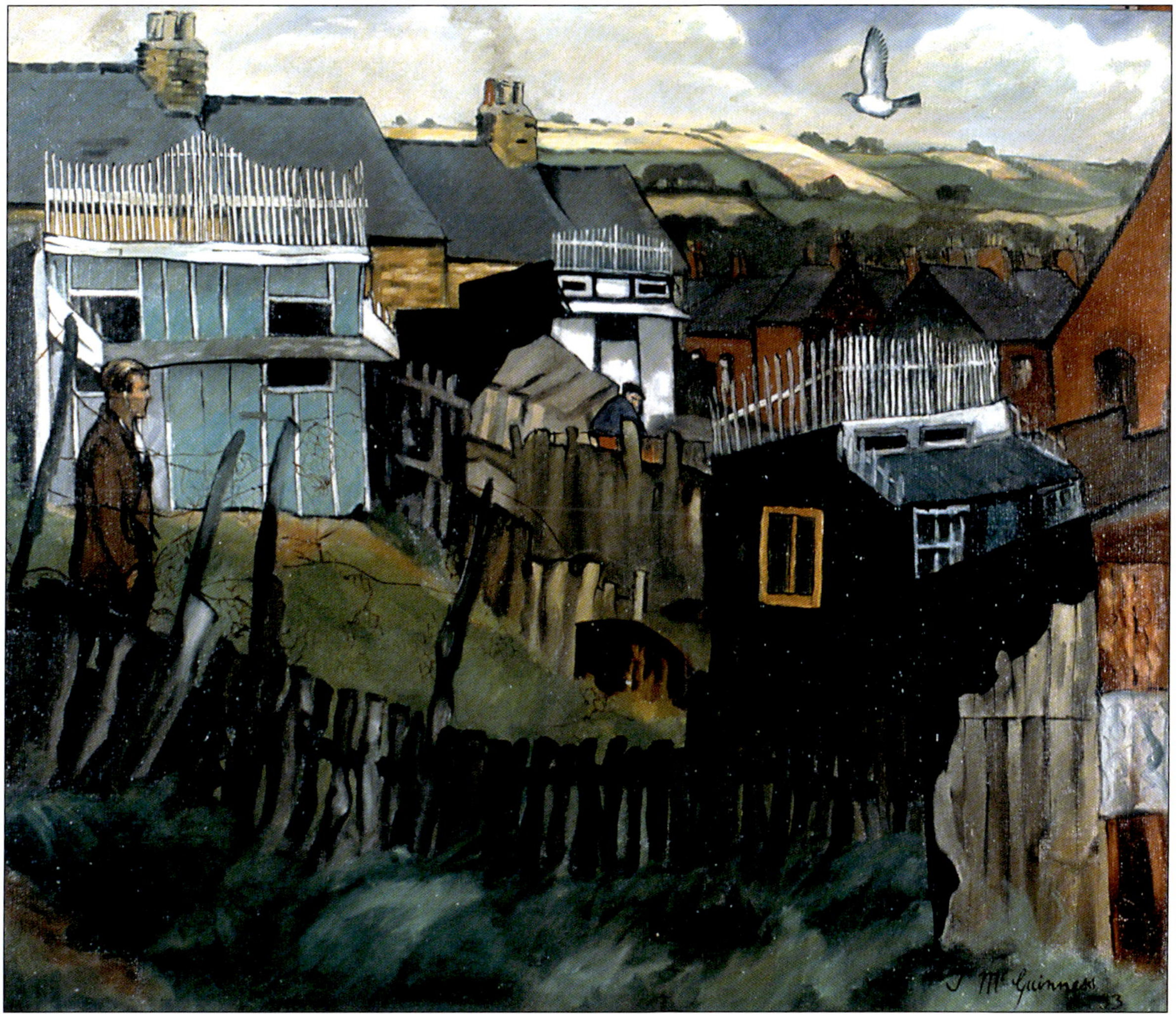

Plate 6

In the Pub 1950s

Unglazed oil on strawboard - 63 x 75cm

Painted in six hours and based on a drawing done in situ in a pub near Rossi's Corner, Bishop Auckland, Tom was attracted to the scene as it reminded him of Walter Sickert's 'Ennui' - boredom. Tom assumed the couple to be married and uses a low vantage point and acute perspective to exaggerate the mood. He used strawboard, a material Tom knew to have been used by Lautrec, as the colour of board meant areas could be left blank. He used size rather than primer on the surface to save paint. Folk singer Jez Lowe used this image for the cover of his CD 'Tenterhooks'.

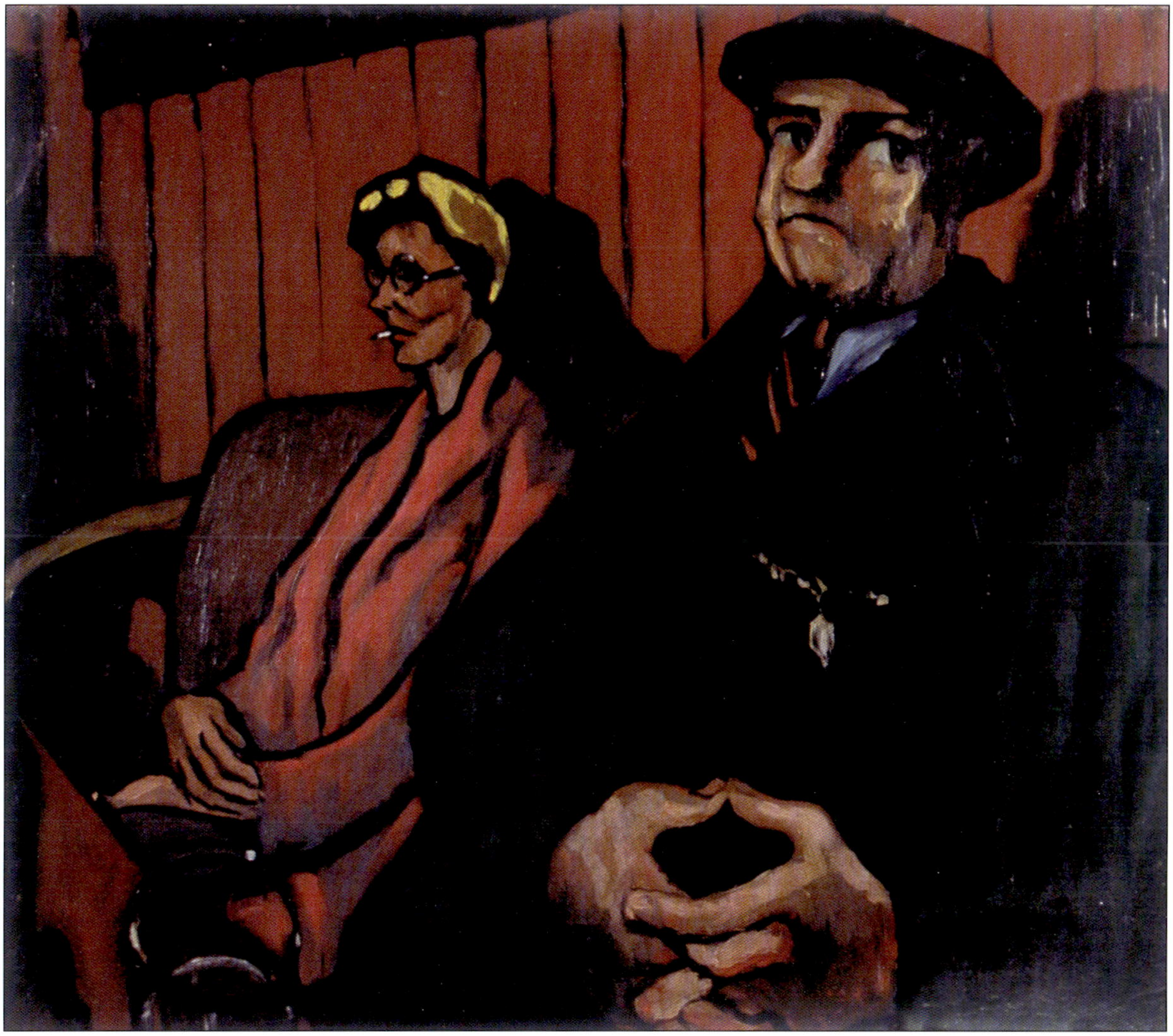

Plate 7

The Filler 1955

Unglazed oil 90 x 98cm

Vibrant use of colour, accentuated by heavy black outlines, typified Tom's work in the early 1950s. Although described as unglazed there is evidence of some use of glaze on the filler's battery pack. Tom was working as a filler at Shildon Drift when this painting was executed. There is increasing evidence of distortion and the artist adds power to the image by placing the viewer at ground level.

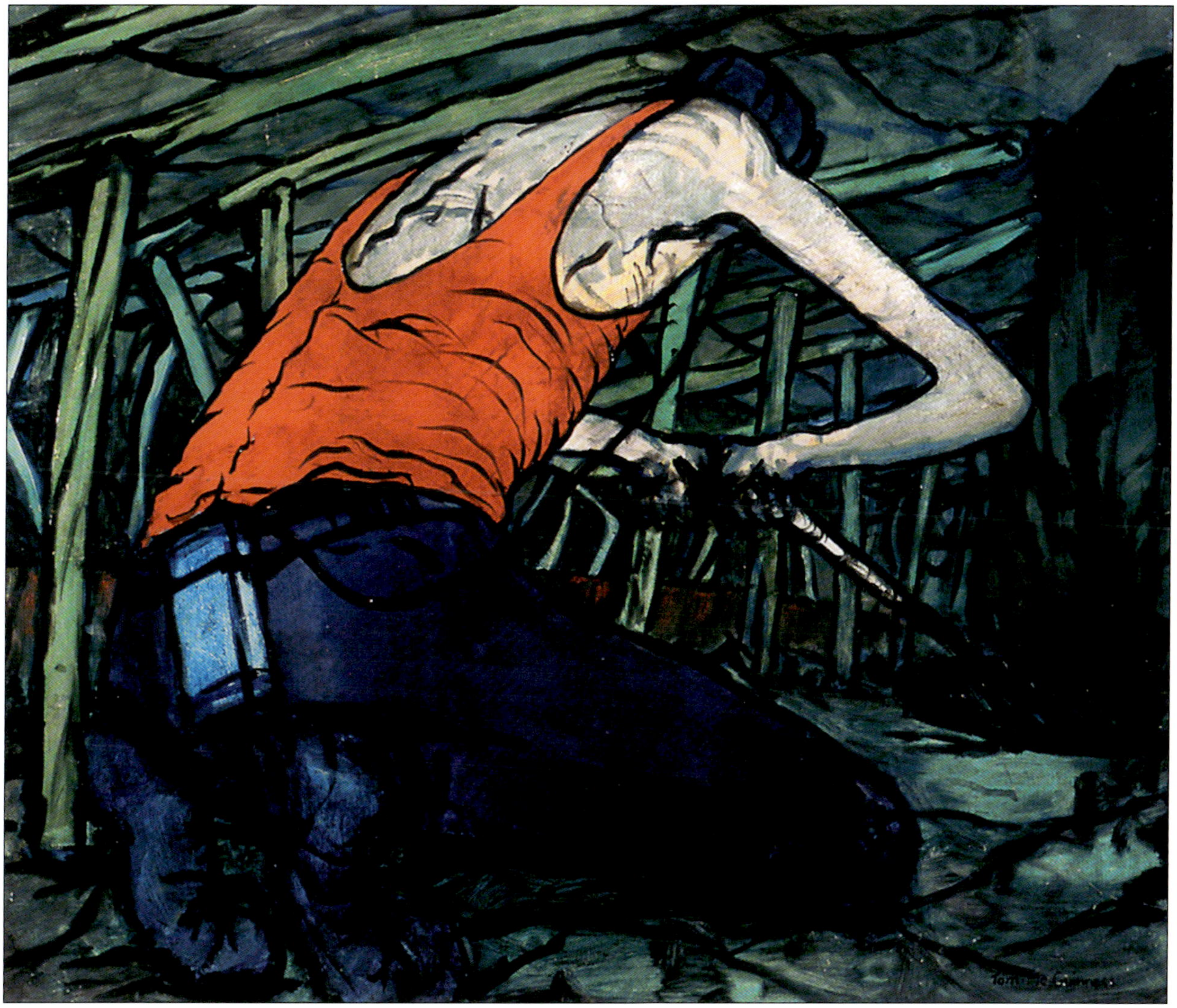

Plate 8

Filler on the Face 1958

Wash, chalk and indian ink on brown paper - 30 x 40cm

In this powerful image, part of the Jack Reading Bequest, the stylistic influence of central European socio-heroic art and the work of Henry Moore can be seen. The bow-shape of the miner appears to have been designed to fit the constraints of the tunnel. This man of steel is master of his environment.

See also page 25

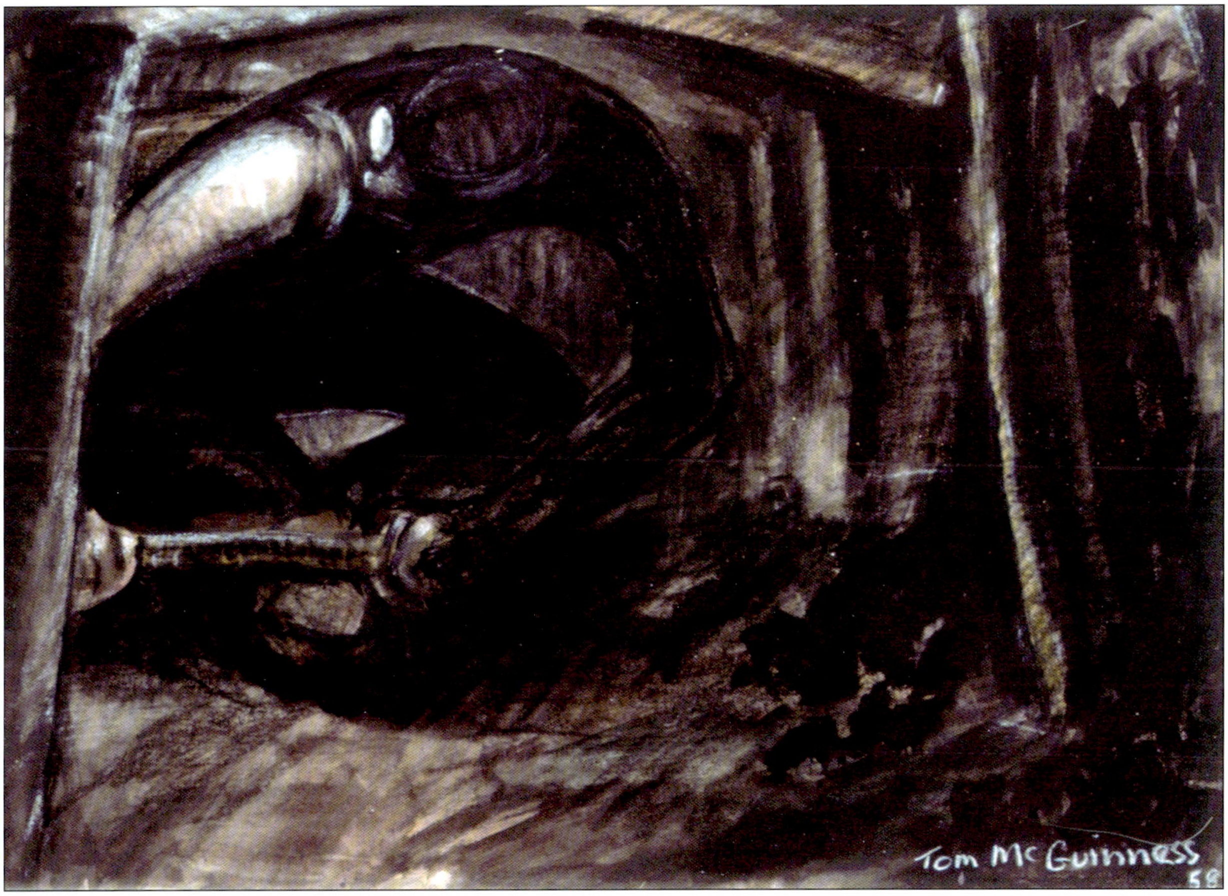
Tom McGuinness
58

Plate 9

Portrait of a Miner 1959

Oil on canvas - 75 x 60cm

Courtesy of Middlesbrough Art Gallery

This portrait of Michael Fairish, in Middlesbrough Art Gallery's current collection, was purchased, together with 'Early Morning Shift', in 1964 following Tom's major solo exhibition at the gallery. On the recommendation of curator Ken Cozens the gallery went on to buy five more works in 1966. Painted from life in his Aunt Winnie's house in Thompson Street, Witton Park, the portrait is of historical interest as it shows Michael wearing a soft canvas helmet for his work with Tom in Brown's Pit, a drift mine at Etherley Dene, Bishop Auckland.

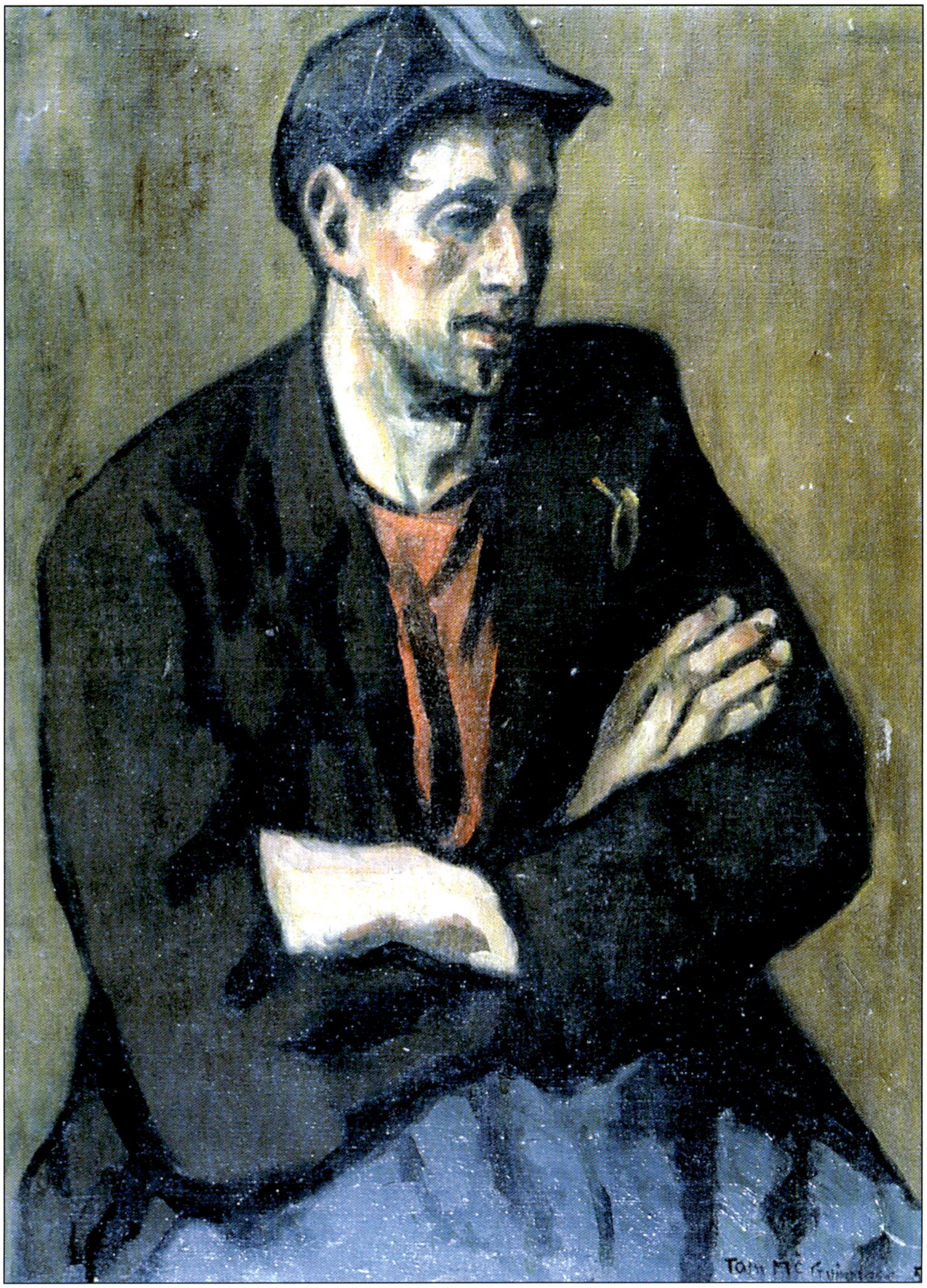

Plate 10

Early Morning Shift 1959

Glazed oil on canvas - 120 x 120cm

Courtesy of Middlesbrough Art Gallery

An early example of a glazed oil, it was based on an monochrome oil sketch executed in the early 1950s. Although inspired by an original scene at Eldon Lane, Tom used his artistic imagination in his representation of the coal mine, leaving the viewer several possible interpretations of the picture including why he chose this particular vantage point.

See also pages 11 & 23

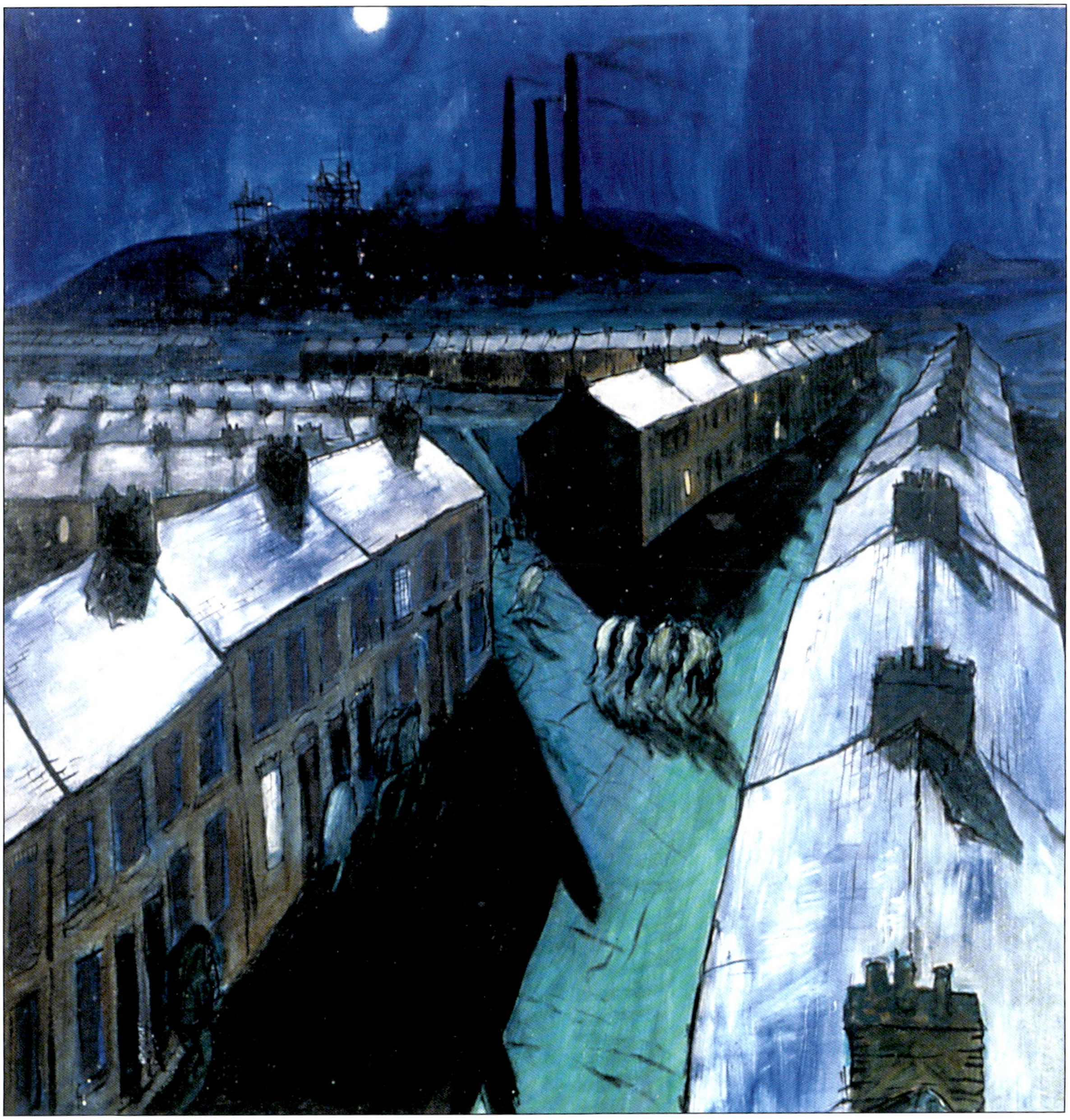

Plate 11

Tailgate 1961

Glazed oil - 45 x 45cm

This almost abstract glazed oil, painted during Tom's 'green period', uses glazes to glorious effect. To help achieve this he has used, amongst other pigments, veridian and ultramarine which have a natural lucency, enhancing the intangibility of the image. Tom has deliberately distorted the disused railway to give the impression of movement within the tunnel "...the floors coming up, the sides coming in" and the men appearing to be swallowed up whole by the leviathan-like mine - all that remains are the tiny stabs of light from their cap lamps.

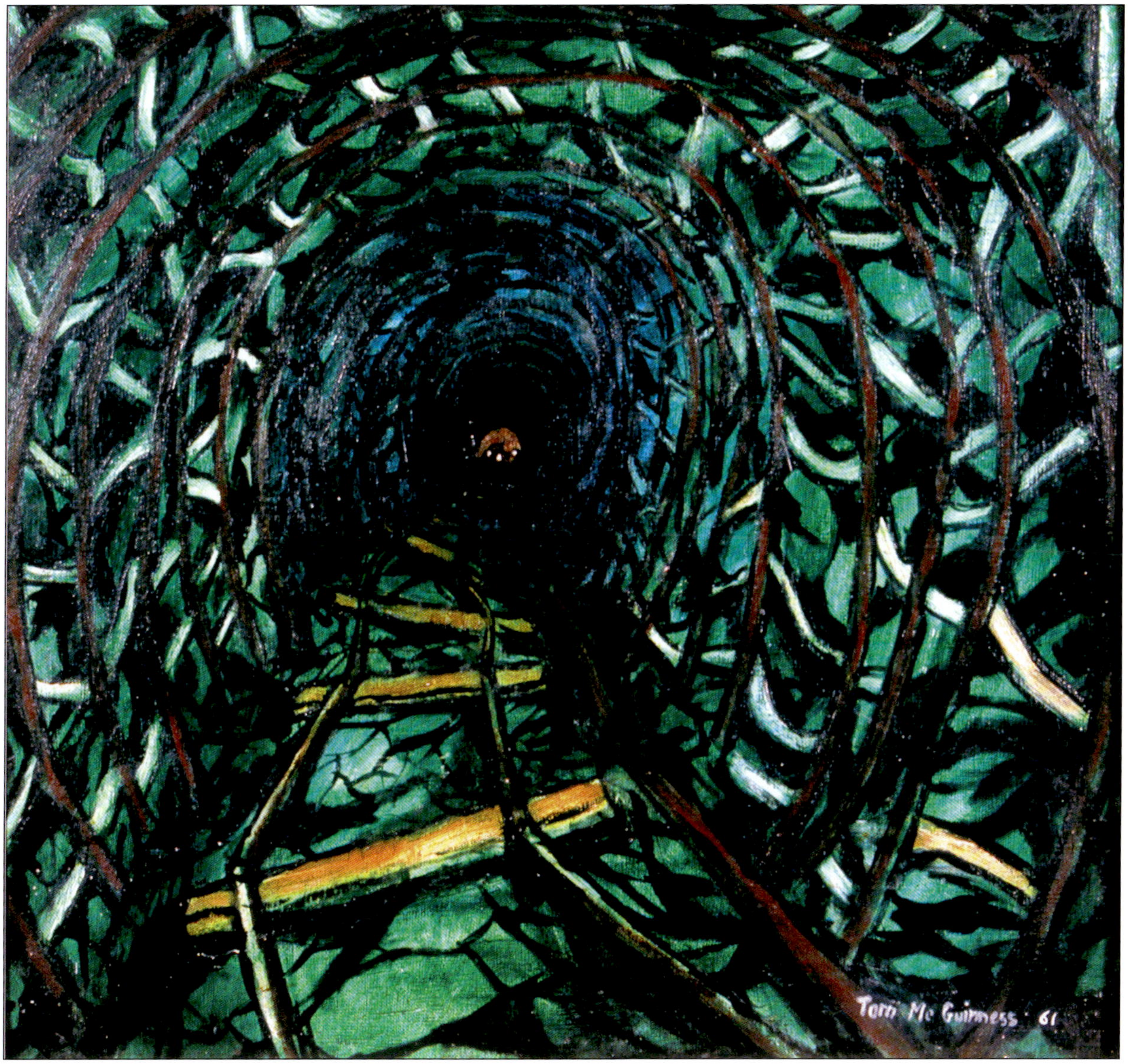
Tom McGuinness 61

Plate 12

Men in the Cage 1961

Glazed oil on canvas - 80 x 60cm

Painted following a cage accident at Kelloe Colliery, the cold, tired miners in Tom's painting return to the surface following their day's hard toil. They squat on their hunkers to avoid sitting on sharp objects and their weary faces turn towards one another for comfort and comradeship. They can be seen to represent man's continuing confinement, oppression and subjugation - a universally recurring theme in all great art.

See also page 17

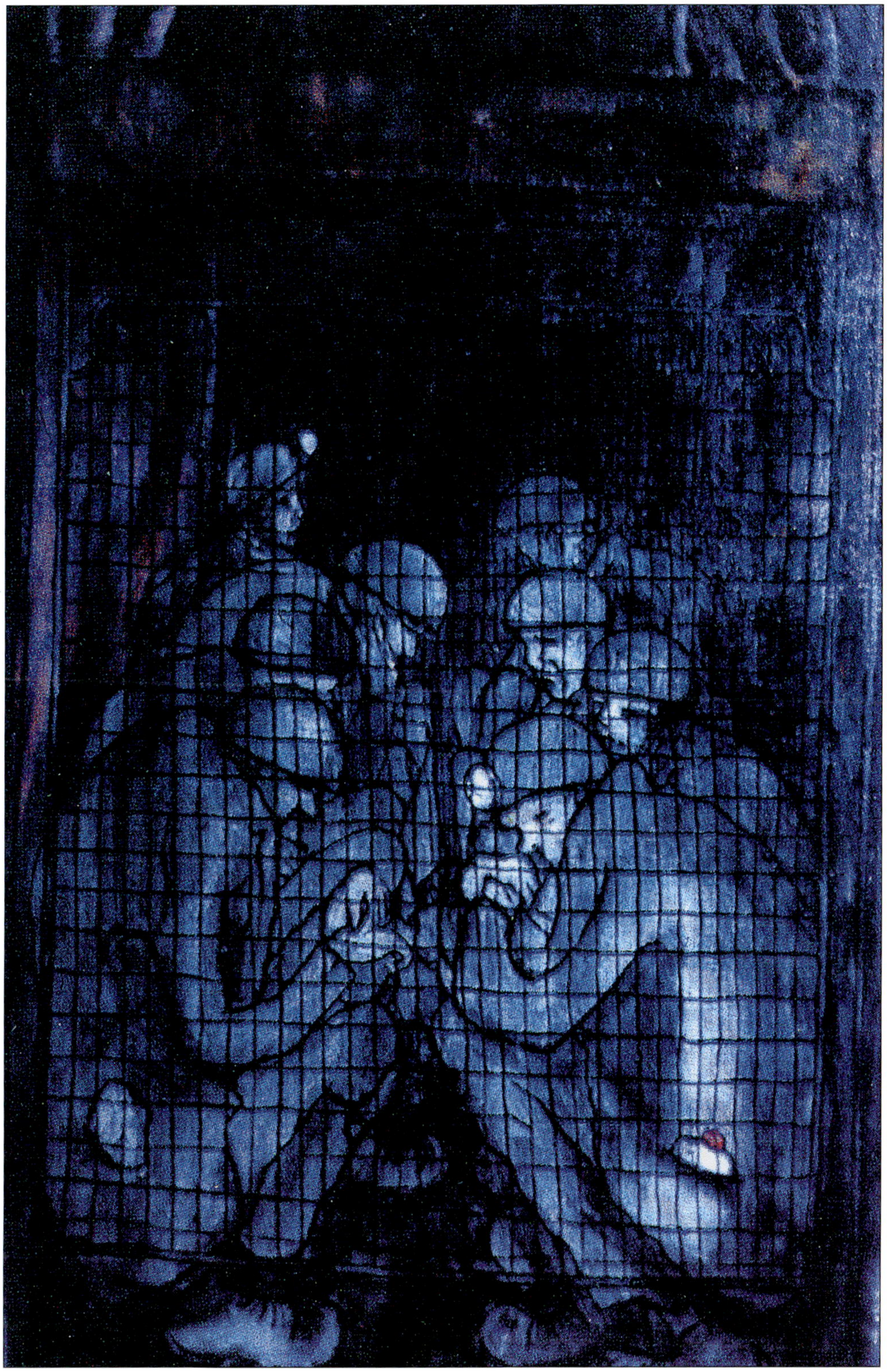

Plate 13

Miner on the Face c1963

Egg tempera on coloured paper - 60 x 90cm

Tom uses tempera for the first time and to great effect. This nightmarish image of the entombed miner almost screams out at the viewer in its Kafkaesque vision of the trapped victim of the oppressive goaf. His tortured arms, bent in homage to the breaking props, have almost metamorphosed into the limbs of an insect or of a bird. Tom accentuates the ultimate in claustrophobia by making the miner appear too big for the painting as the constricting frame of the picture is crushing him to death.

See also pages 25 & 26

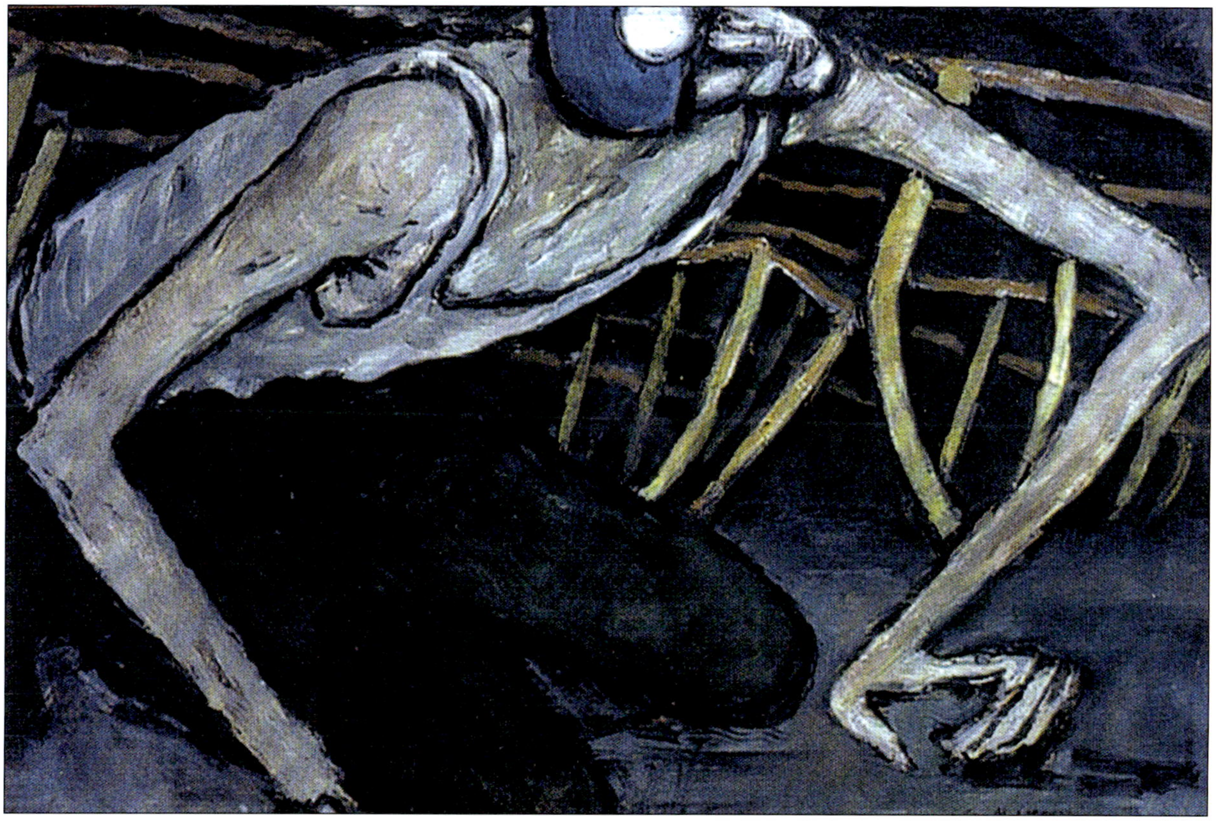

Plate 14

The Potato Pickers 1960s

Oil - 90 x 120cm

A particular favourite of Tom's, he requested that this painting be included in the book for its nostalgia value. With resonances of a Van Gogh subject, Tom's heavily distorted depiction shows the women pickers putting potatoes into the pouch formed by folding their aprons. Tom had spent his October 'potato picking holidays' doing just this. It was back-breaking work and by use of distortion, Tom by implication, is drawing an analogy with the rigours and toil of the miners.

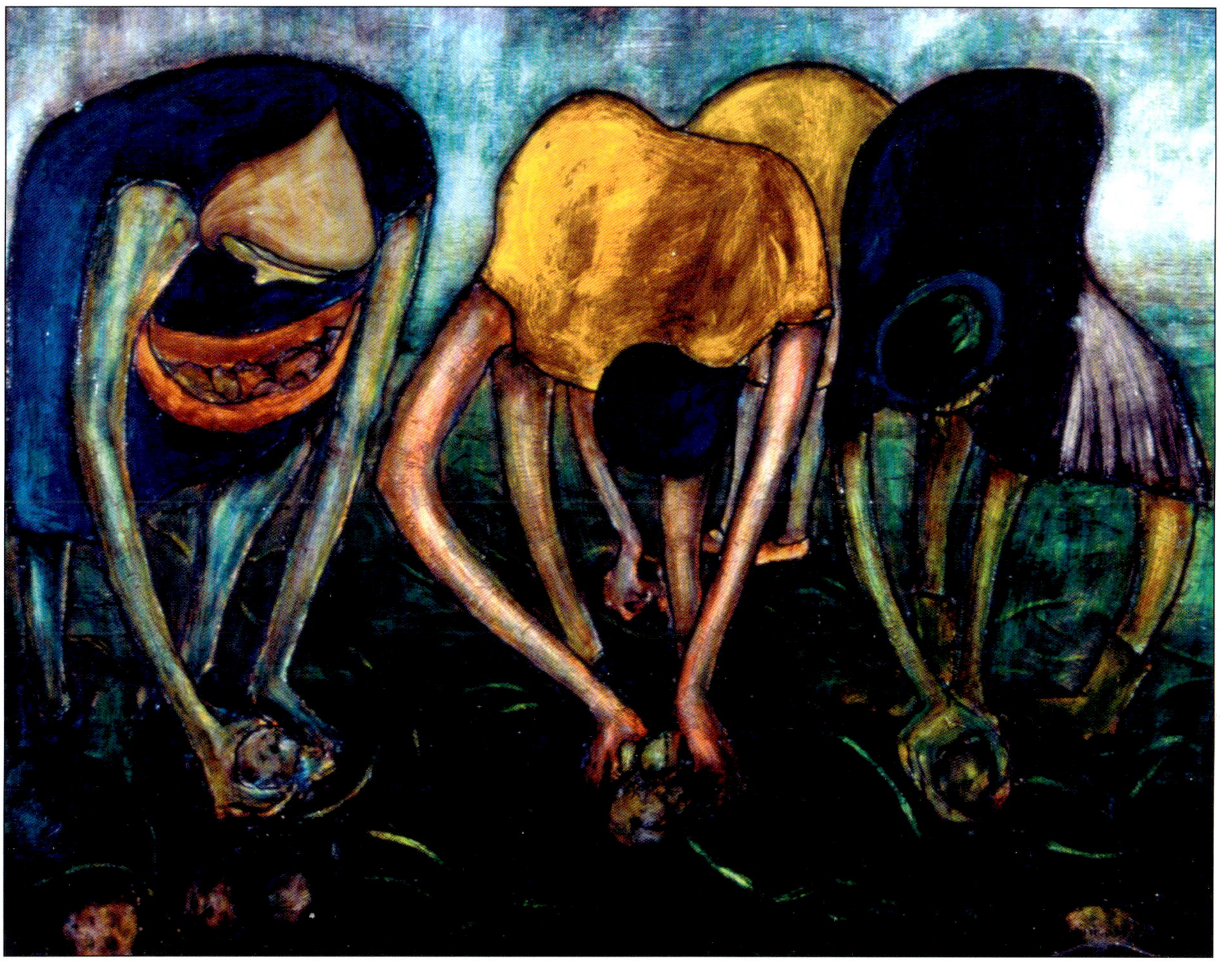

Plate 15

Bait Time 1963

Glazed oil - 30 x 150cm

Bought by Sid Lockey, a close friend, this painting has been compared in composition with Leonardo's 'Last Supper'. As Catholics, this would have resonance to both Tom and Sid. Tom's simple use of glazes creates an almost stained-glass window effect in raising the figures out from the background.

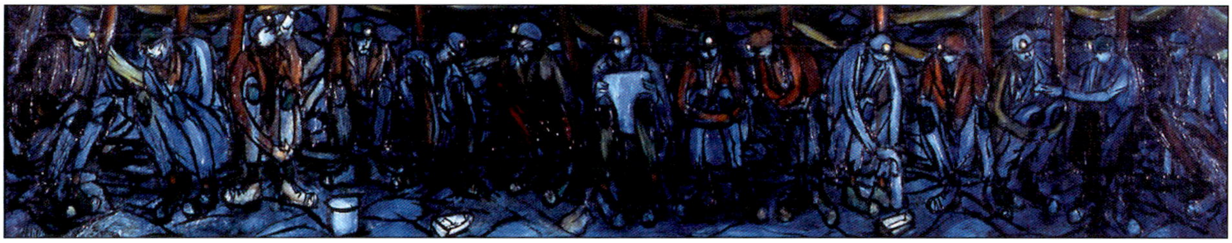

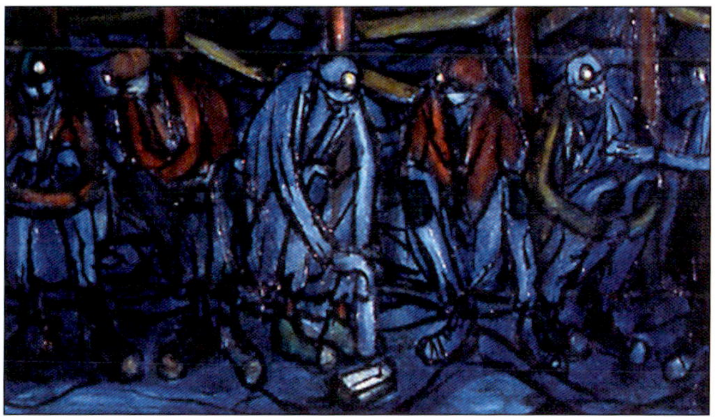

detail

Plate 16

Beach Scene 1970s

Mixed media- 30 x 90cm

In response to his then agent Lord Hirshfield's suggestion that southern audiences preferred outdoor scenes in brighter colours to industrial subjects, Tom produced a number of beach scenes using his wash-off technique. Despite the change of subject matter, each scene still contained geometrical lines - this time the supporting legs of the piers rather than those of the gantries of the pit winding gear. Realising that the subject matter lacked honesty, he returned to his first love, the mines.

See also page 16

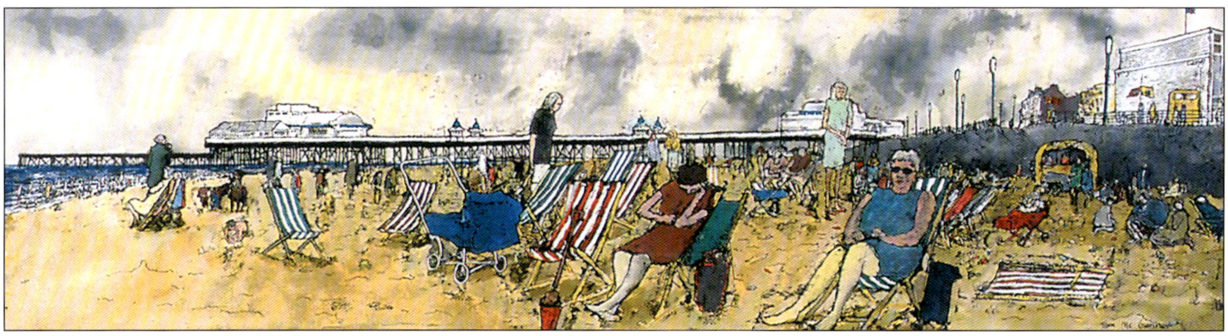

Plate 17

The Pit Road 1971

Etching - 8 x 12cm

An early etching, Tom's potential as a printmaker is already obvious in the complexity of the design for the Pit Road and the precision with which it is executed. Unusually, we see women and children filing silently towards the colliery gates - awaiting news of a disaster perhaps? Tom was to refine his intaglio printing technique by mastering drypoint, engraving and aquatint.

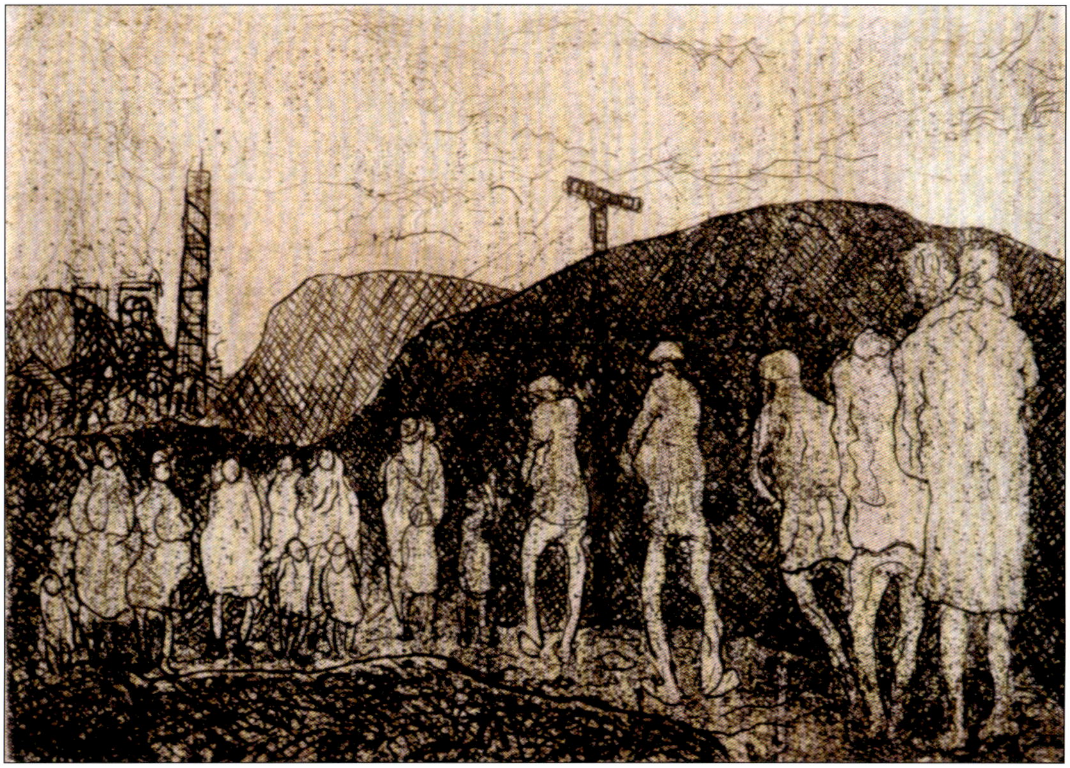

Plate 18

Stone Men 1975

Glazed oil, canvas on panel - 50 x 75cm

One of the most dangerous jobs in the pit, the stone men are bringing down the loose stone from the roof after the shots have been fired prior to the supporting rings and planking being placed to form the roadway. Tom's masterful contrast of tones creates an internal light source to illuminate the scene. This narrative painting relies more on colour and composition than on distortion to gain its powerful effect.

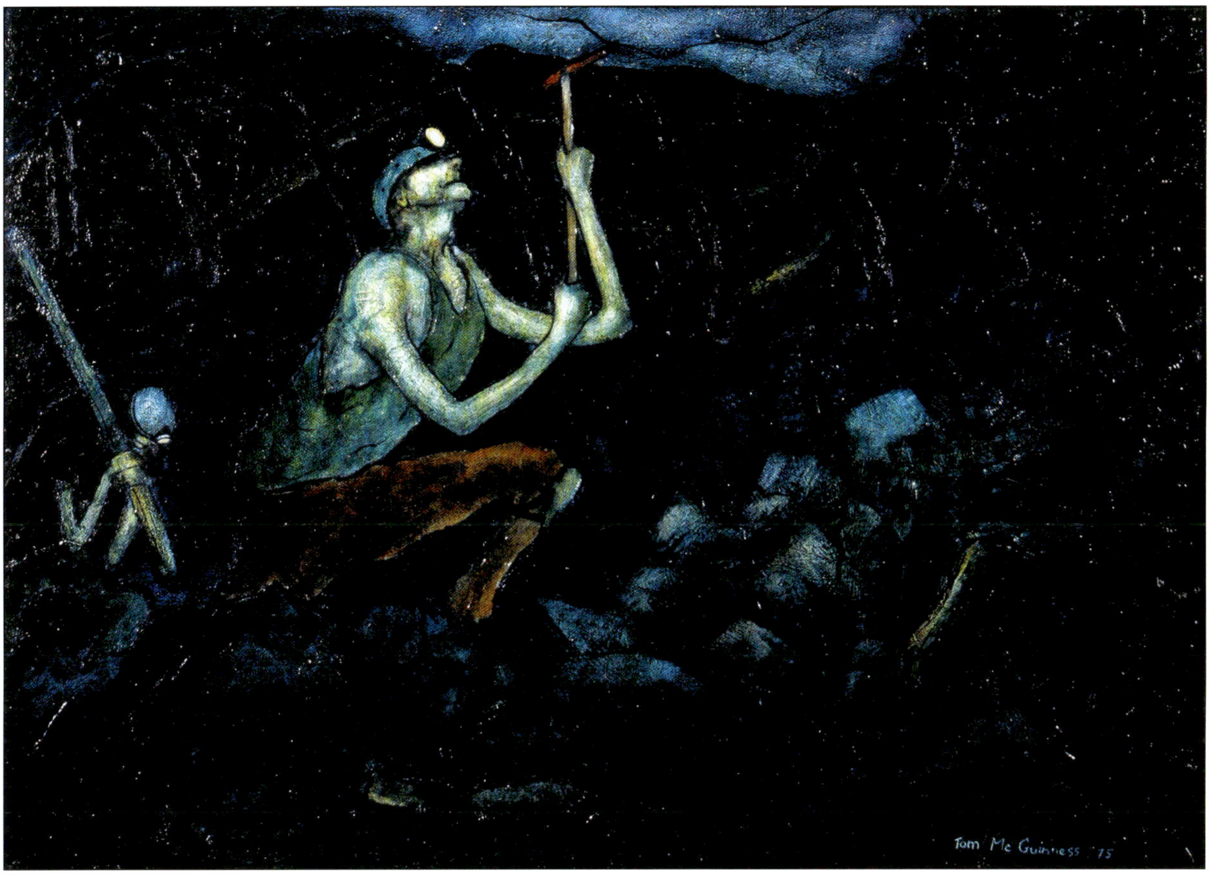
Tom Mc Guinness '75

Plate 19

Miners' Gala 1976

Oil on canvas -120 x 94cm

Commissioned by Barclays Bank at the suggestion of Harry Ramsdale, owner of Gallery 2 in Darlington where Tom had previously exhibited, this large oil still hangs today in the Durham Market Place branch of the bank. It is most apposite as Barclays have been long-term sponsors of the Gala. Tom's picture with its vibrant primary colours showing the miners at play contrasts starkly with the sombre hues of his working pitmen.

See also page 16

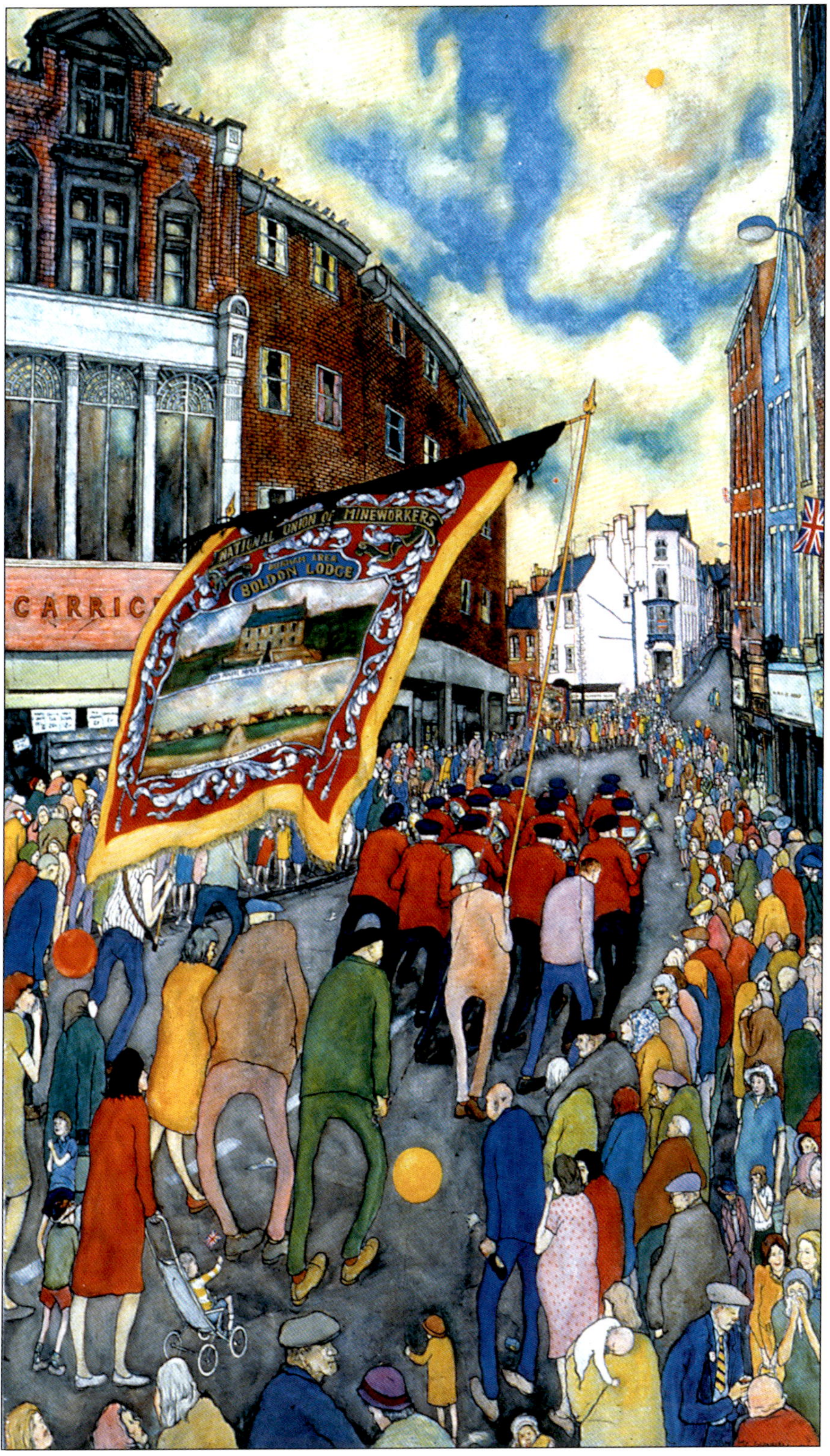
NATIONAL UNION OF MINEWORKERS
BOLDON LODGE
GARRIC

Plate 20

Night Shift No. 2 1977

Aquatint etching - 20 x 13cm

Inspired by the prints of Goya, Tom uses aquatint to add strong tonal qualities to his etchings in order to produce dramatic contrasts of light and dark. Tone was such a vital element in the pit where light was at a premium. Here Tom's trademark moon presides over the miners ascending the gantry prior to their descent into darkness.

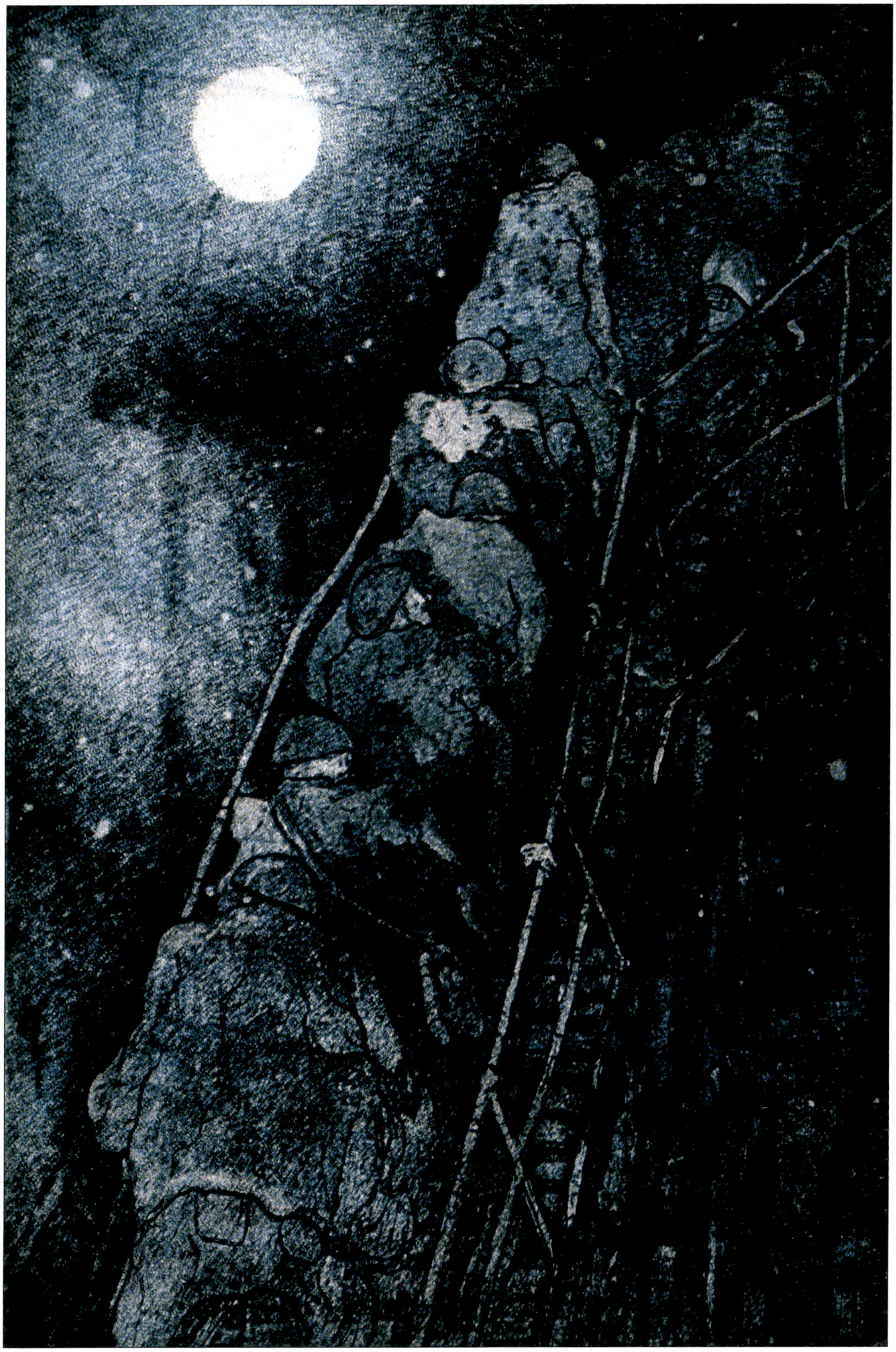

Plate 21

Harveymen Going Outbye 1977

Lithograph - 50 x 35cm

This lithograph, together with its counterpart Busty Men Going Inbye (see frontispiece), was produced at Charlotte Press, Newcastle, after Tom had been awarded an Arts Council bursary to study the technique of lithography at Lowick in Cumbria in 1976. The joyful men, with shift finished, appear to be dancing their way home. The lyricism inherent in the work is created by the low vantage point and the plunging curves of the tunnel.

See also page 18

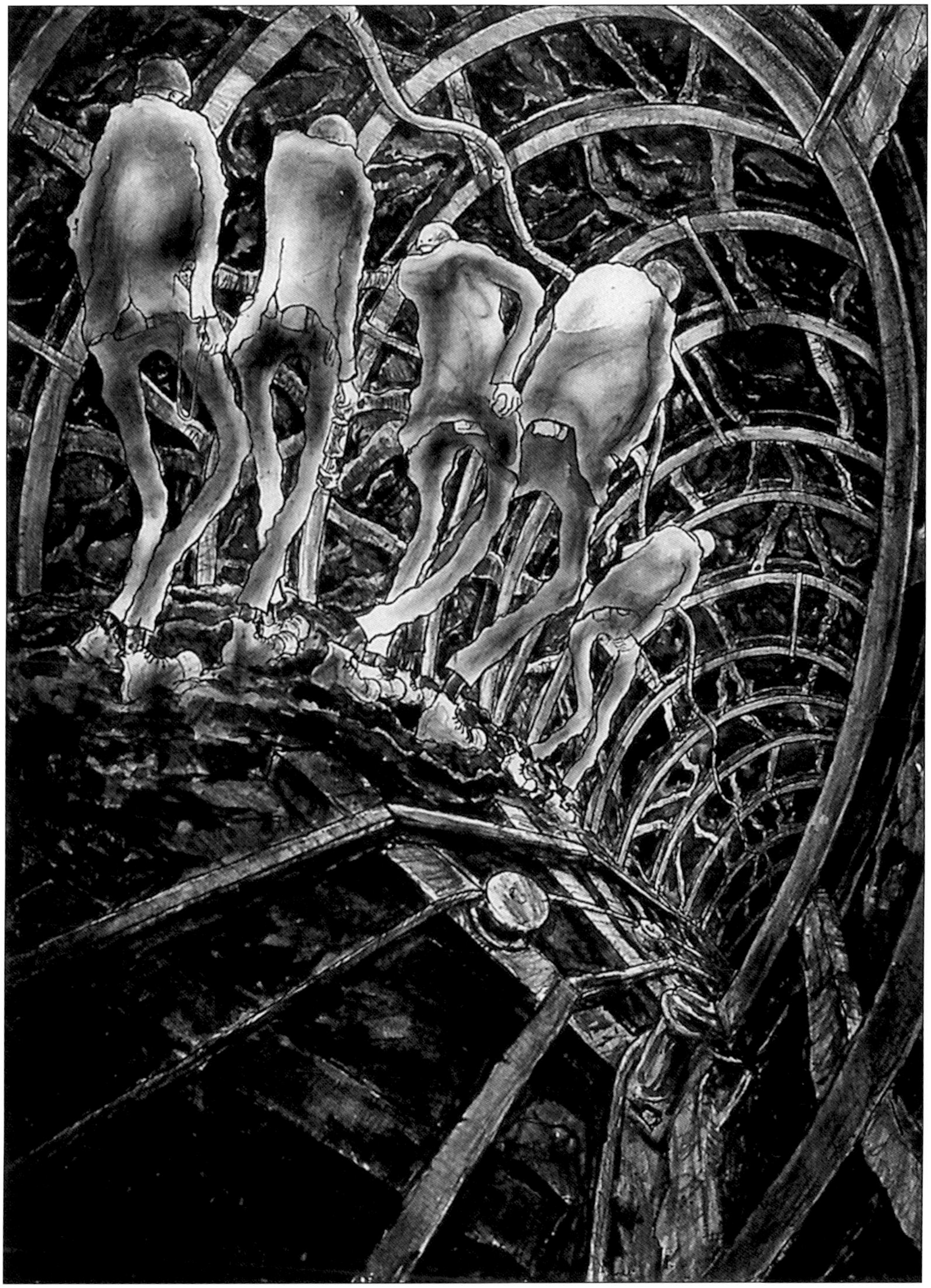

Plate 22

Going to School 1977

Oil, canvas on board - 60 x 120cm

One of a series of street scenes that Tom painted in the 1970s depicting everyday life in the pit village. All the scenes are based on real places although Tom adapted them where necessary. In 1970 Tom, having sold a painting for £120, took six weeks off work, touring Durham in order to draw the local pit communities. In almost all of these paintings the colliery retains guardianship of the scene.

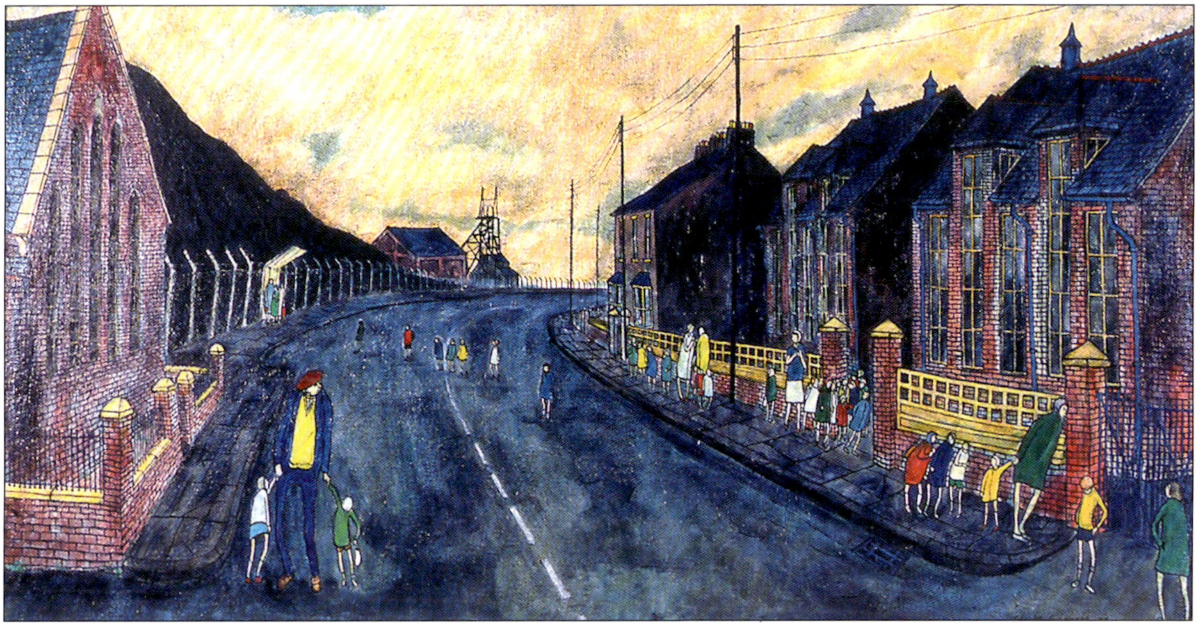

Plate 23

The Gangway 1979

Engraving and aquatint - 30 x 20cm

Rembrandt, one of Tom's favourite artists, combined the techniques of the etcher - the printer using acids to erode the plate, with that of the engraver, the armourer - using a burin to score the surface. Tom emulated him to great effect in this superb example of the printer's craft.

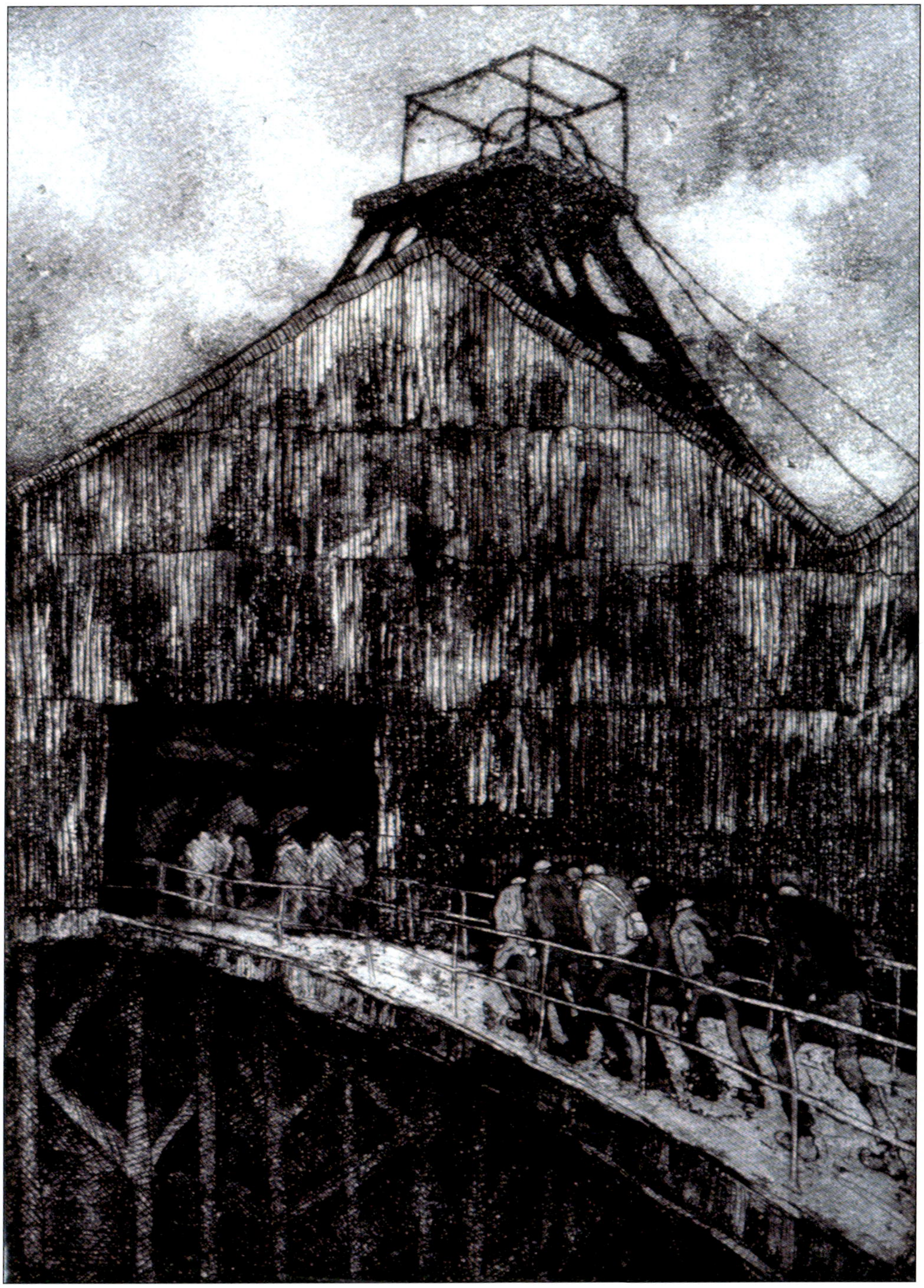

Plate 24

Harvey Manrider 1979

Coloured lithograph - 50 x 60cm

This supreme example of the lithographer's craft is notable for its technical quality and for its subject matter. The freedom of design offered by lithography is constrained by the necessity to overprint the image four times to create the final coloured print. Here Tom's colleagues are going home on the man-riding conveyor belt, exhausted after their shift, or are they being conveyed towards an unknown fate in an almost biblical scene reminiscent of Jonah being devoured by the whale?

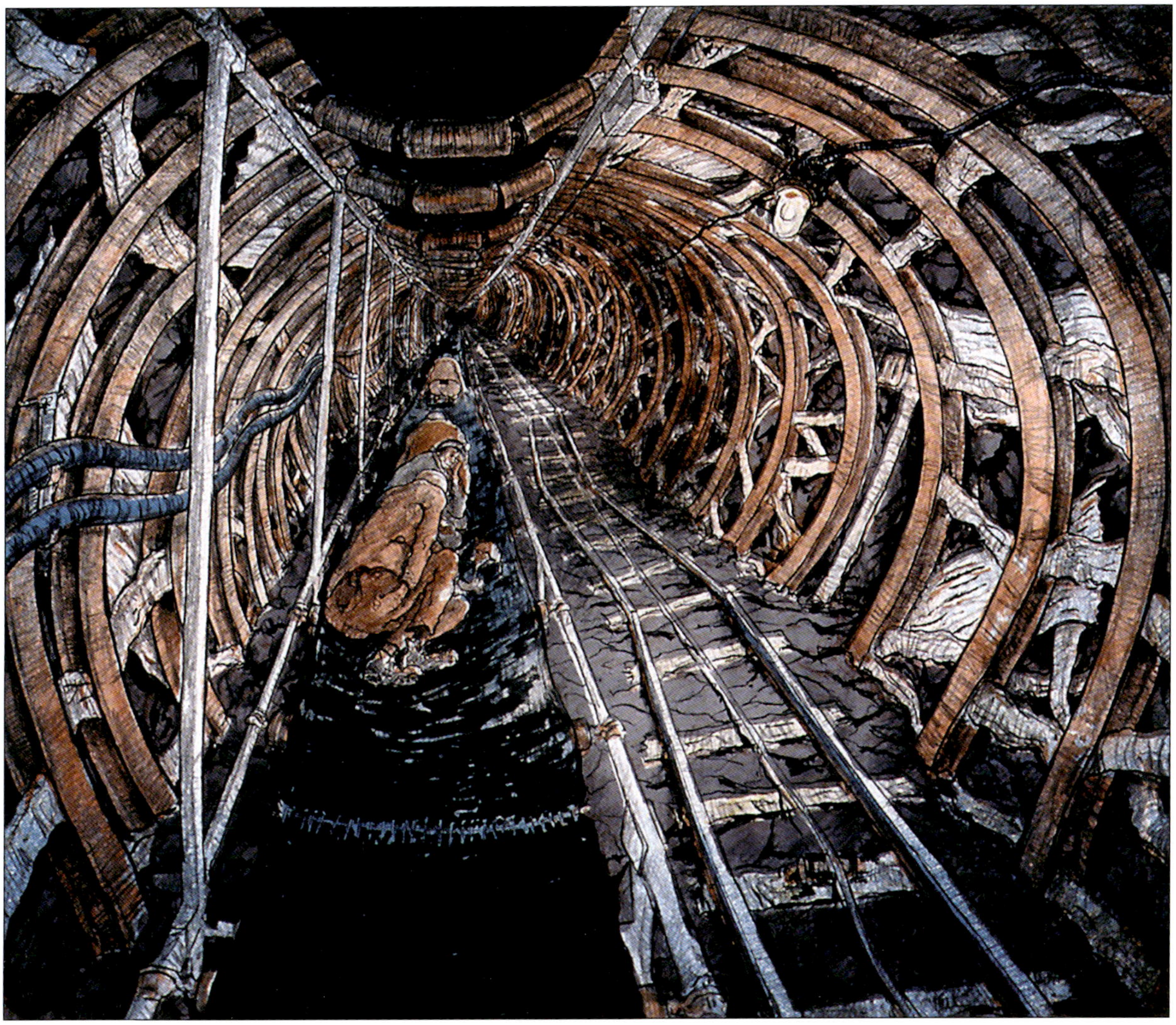

Plate 25

The Drawers 1982

Coloured stone lithograph - 53 x 68cm

In the cold eerie blue light the drawers perform their well choreographed dance as they dismantle the redundant ring girders which removes the tension from the roof. This extremely dangerous procedure required a high degree of synchronicity and physical co-ordination. Their balletic gestures mirror the contours of the tunnel and belie the imminent danger inherent in their task.

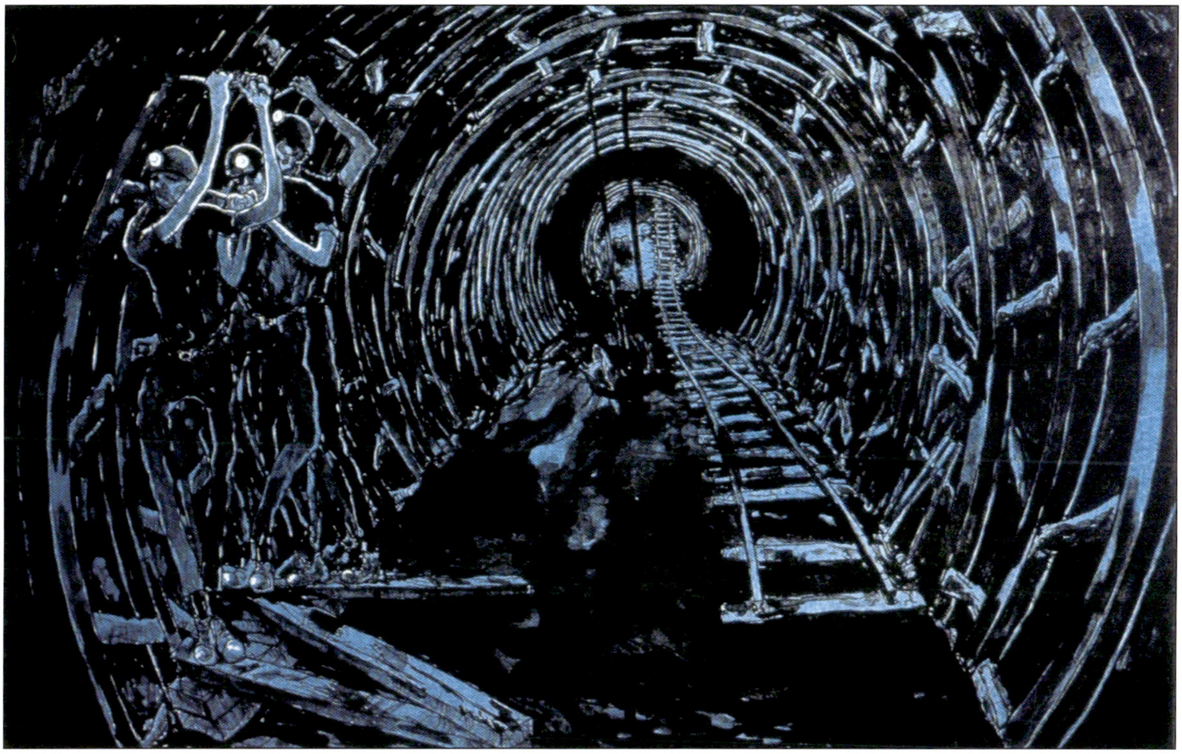

Plate 26

Nightshift 1980s

Gouache and poster block wash-off on cream Ampersand paper - 35 x 25cm

The fairytale quality of this experimental picture, the subject being drawn entirely from imagination, is the result of Tom's attempt to create the effect of a halo round the moon. This picture has a magical feel and is completely different from any of Tom's other wash-offs. It hangs on the wall in his sitting room to this day.

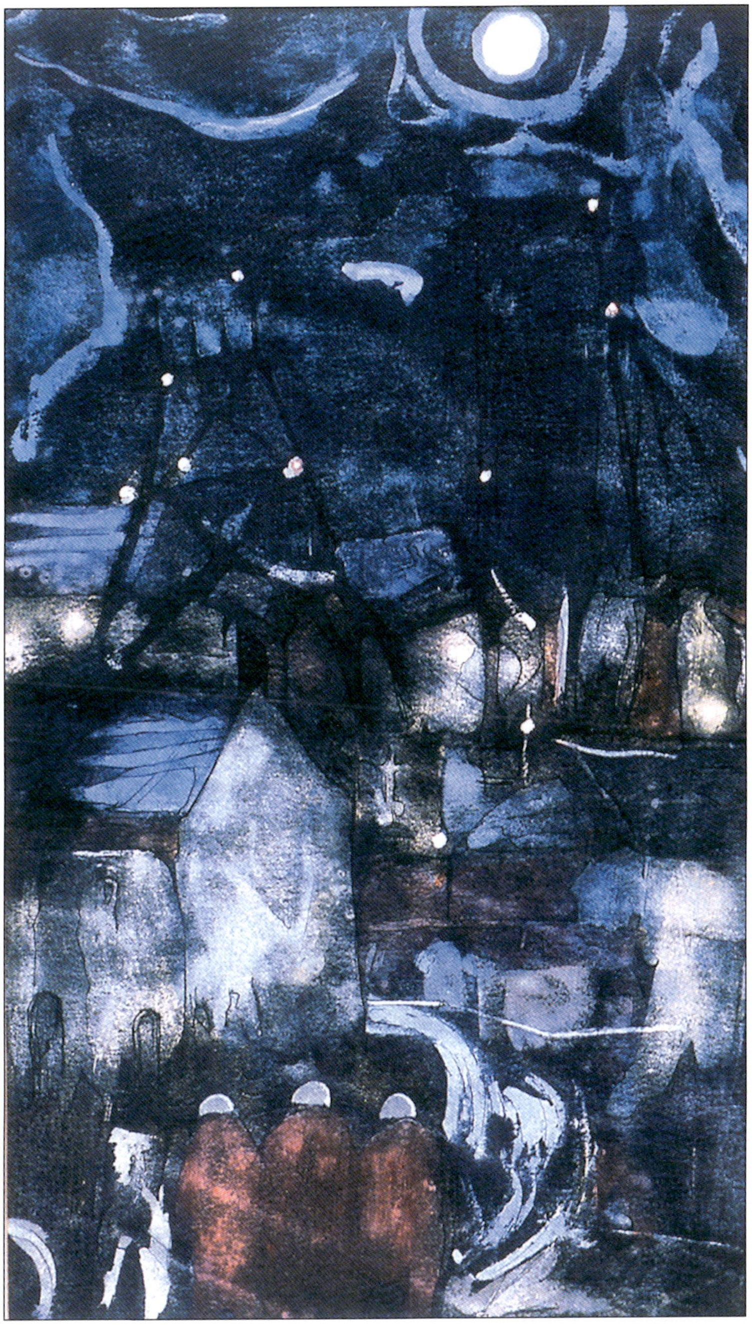

Plate 27

2 a.m. 1981

Etching - 35 x 6cm

This dark image depicts the history of mining in a cross-sectional form. The modern miner is in the lowest seam accessed only by new mining technology whilst his ancient counterpart toils above him in primitive conditions. The overseer, the colliery, anchors them in their respective time whilst the ever-vigilant McGuinness moon presides.

Plate 28

Starting the Cut 1982

Coloured lithograph - 30 x 90cm

This design exemplifies Tom's fascination with machinery and the mechanical aspects of mining and demonstrates his versatility as an artist. The unusual shape of the lithograph, with the impression of a low underground tunnel and the use of skilled perspective contributes to a sense of depth, leaving the viewer with the feeling that one could almost walk into this long wall face.

See also page 21

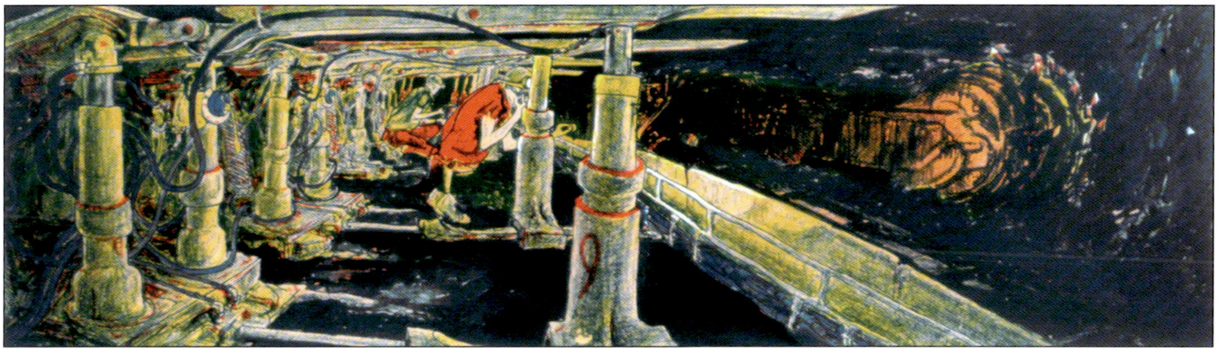

Plate 29

Here Comes the Manager 1982

Oil on board - 28 x 75cm

Painted on the bottom panel of a drawer from an old dresser, this is a recurrent subject for Tom. (He has produced a similar composition in both lithography and graphite). Men were very wary when the manager ventured underground and the news of his arrival spread by means of 'Chinese whispers'. Here we see the sudden appearance of the manager on the left being announced along the line of colliers.

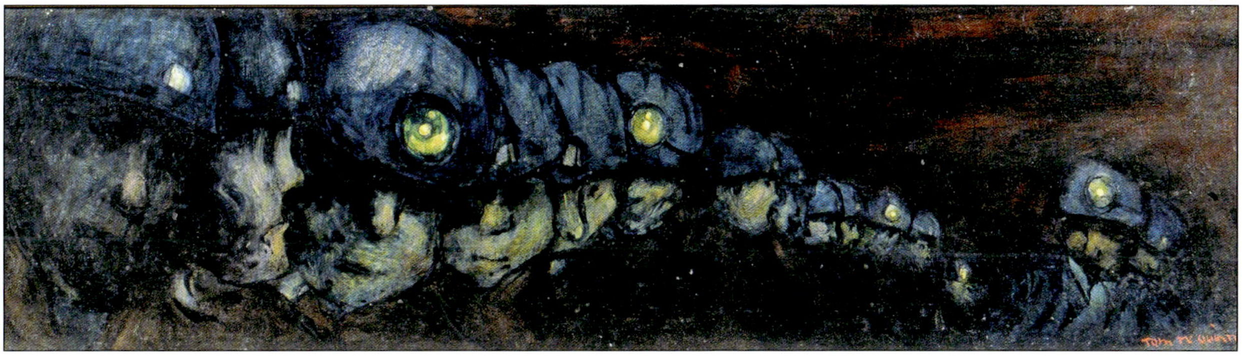

Plate 30

Crucifixion No.1 1980s

Sugarlift etching and aquatint - 15cm diameter

There was a major shift in Tom's subject matter in the early 1980s which coincided with the untimely death of his wife Cathy. In addition to the stained glass window designs for St Mary's Roman Catholic Church in Bishop Auckland, Tom produced a series of religious etchings. In this crucifixion scene, a design reminiscent of the rose window in memory of Father Jarrett - 'Father into thy hands I commend my spirit' - the tortured Christ is viewed from a lofty vantage point, almost from the position of God Himself.

See also reference to Crucifixion series, page 20

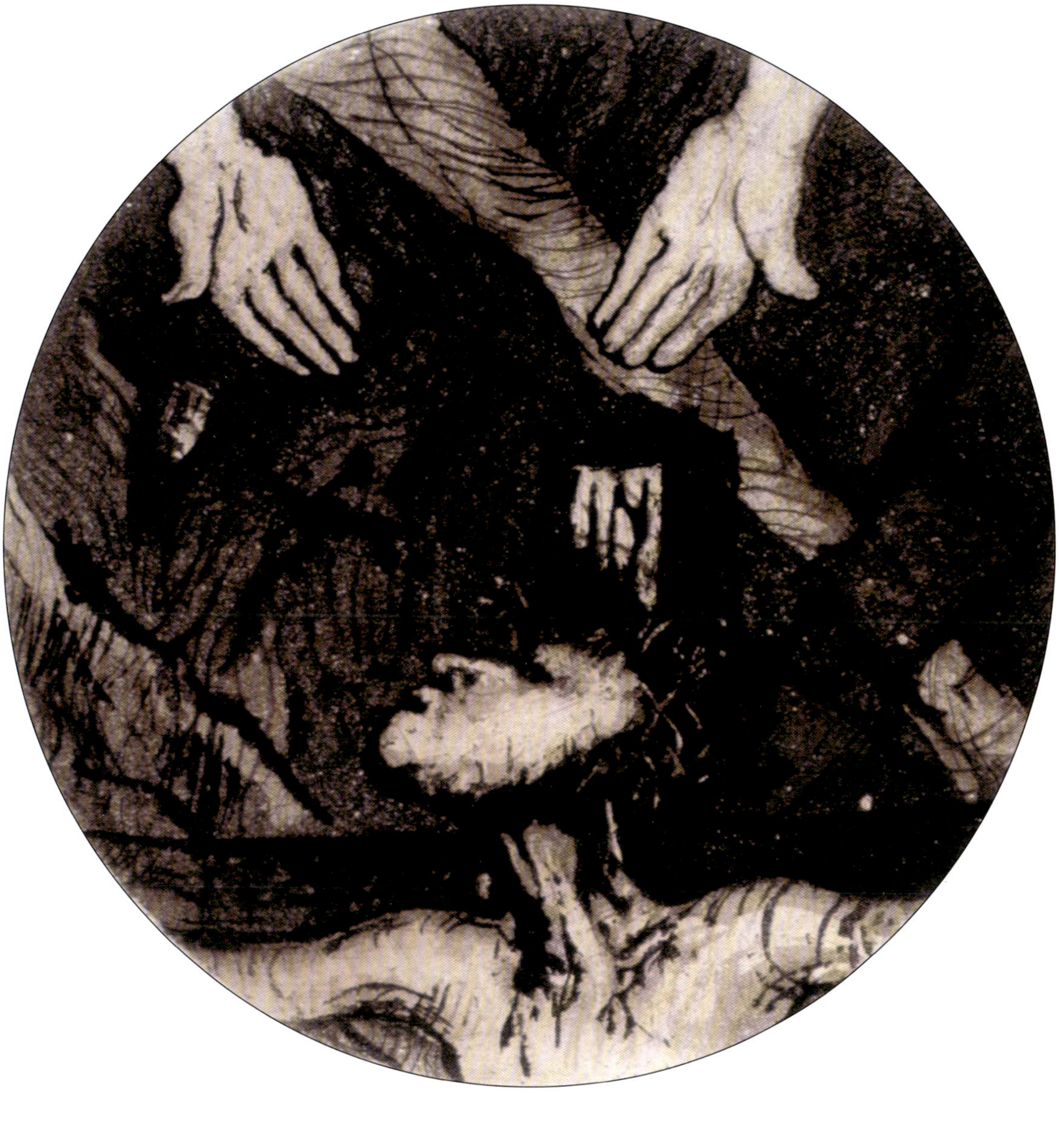

Plate 31

You Must Feed the Starving in Our World 1988

Stained glass window - 185 x 40cm

This is one of eight stained glass windows which were installed in the Blessed Sacrament Chapel in St Mary's Church, Woodhouse Lane, Bishop Auckland. The windows had been commissioned by Father Michael Finnigan, together with a ninth - the sanctuary window - a rose window in memory of Father Jarrett, the founding priest of the church. This profoundly disturbing image was used as an illustration in the textbook, Religion in Focus: Christianity in today's world. The stained glass was crafted by Edna Partridge.

See also stained glass windows, pages 19 & 20

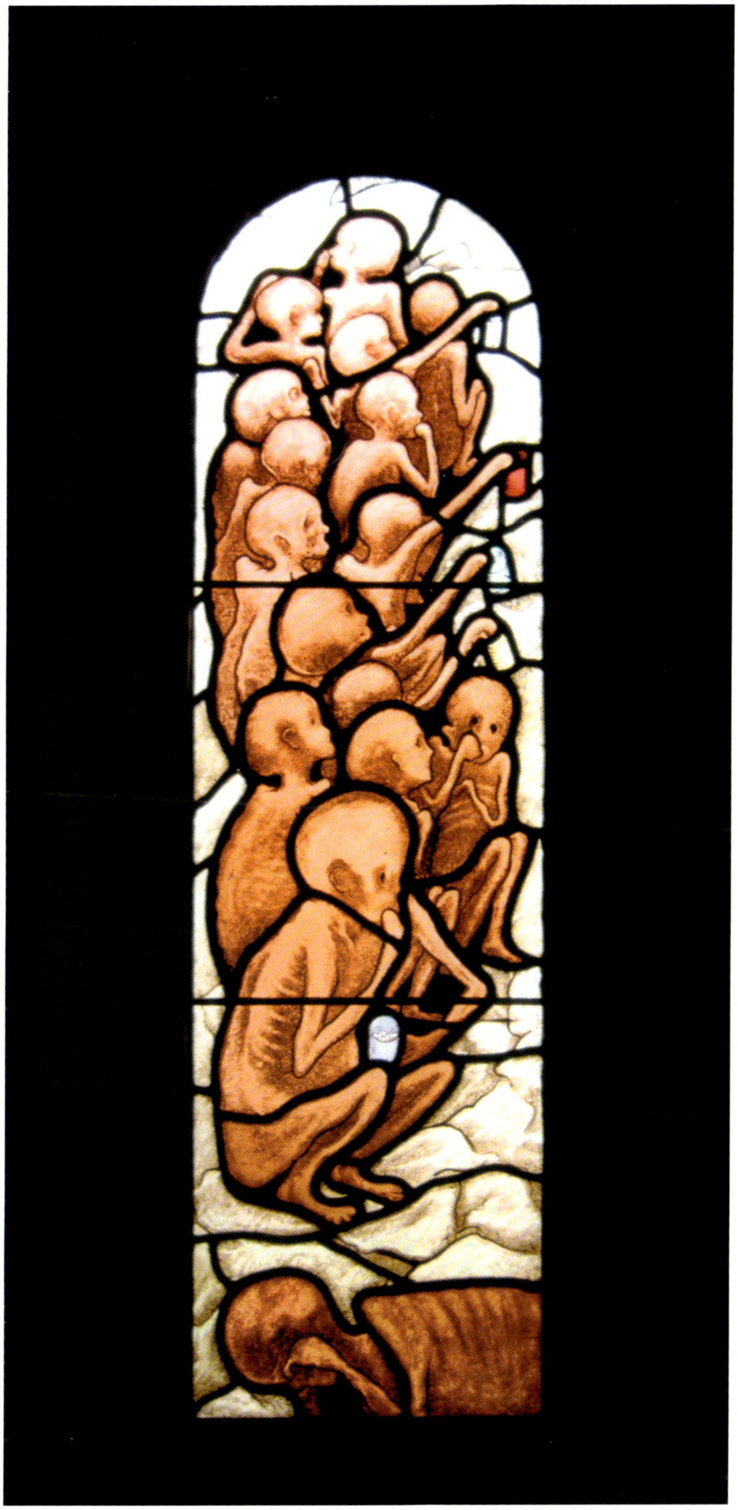

Plate 32

The Agony in the Garden 1985

Oil on canvas - 42 x 42cm

This small oil painting was the intended design for one of two panels for St Mary's Church in Bishop Auckland. Designs of similar dimensions for the tabernacle were used to produce the required four enamel panels - crafted by enamel artist Evan Long. This design was never used.

See also page 20

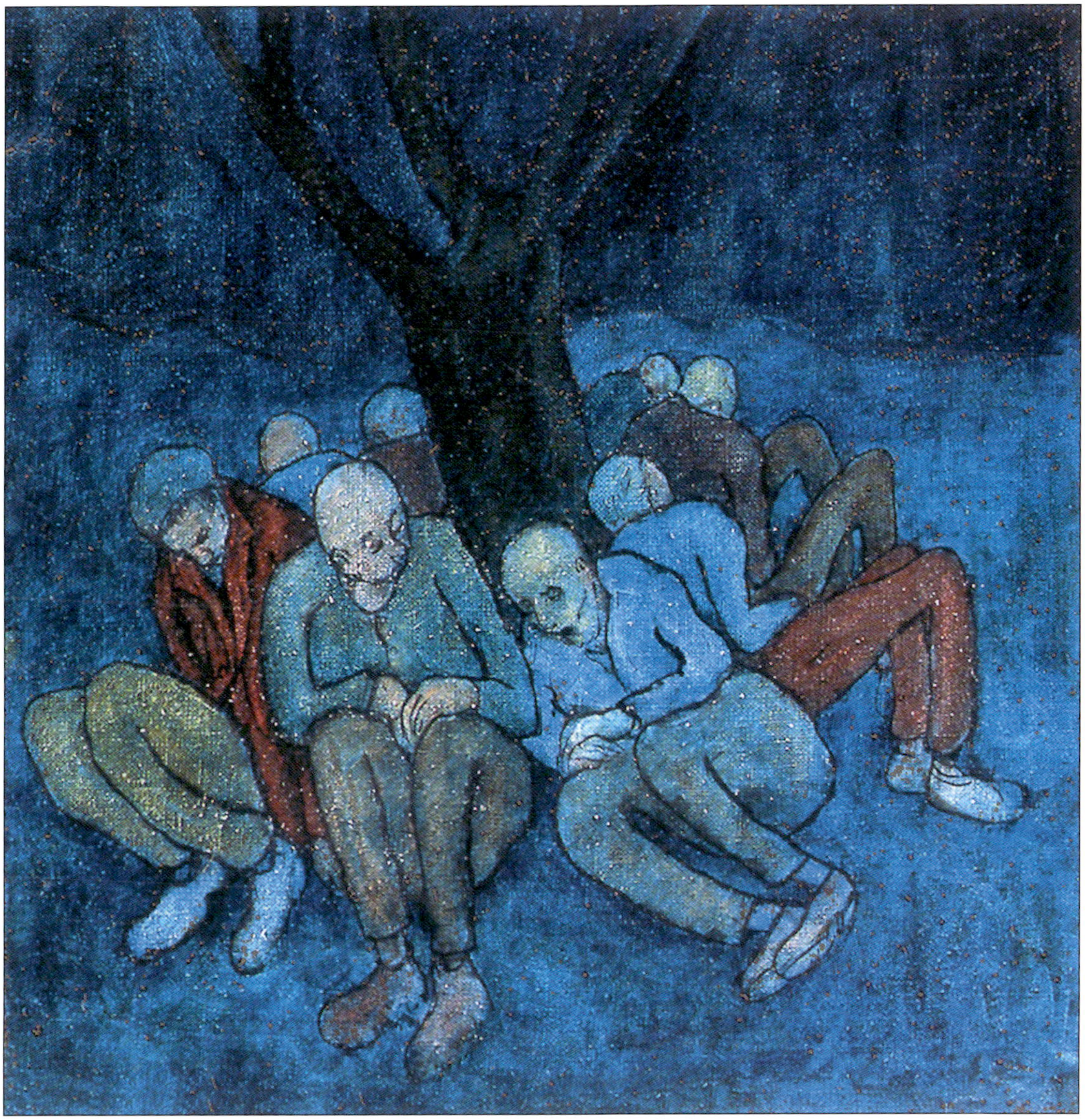

Plate 33

The Power Station 1985

Mixed media on paper - 28 x 28cm

This is one of twenty images produced as the result of a residency at Hartlepool Nuclear Power Station. The paintings, all mixed media wash-offs, continue Tom's divergence from his usual mining theme. This series allowed Tom to move towards abstraction to depict this more fragile, transient world in which the power source is ill-defined leaving an impression that nuclear power, a more nebulous concept than coal, is without definite shape, form or even colour. Line, tone, colour and composition - the four main elements of a painting - are all sacrificed in Tom's interpretation of nuclear fission.

See also page 20

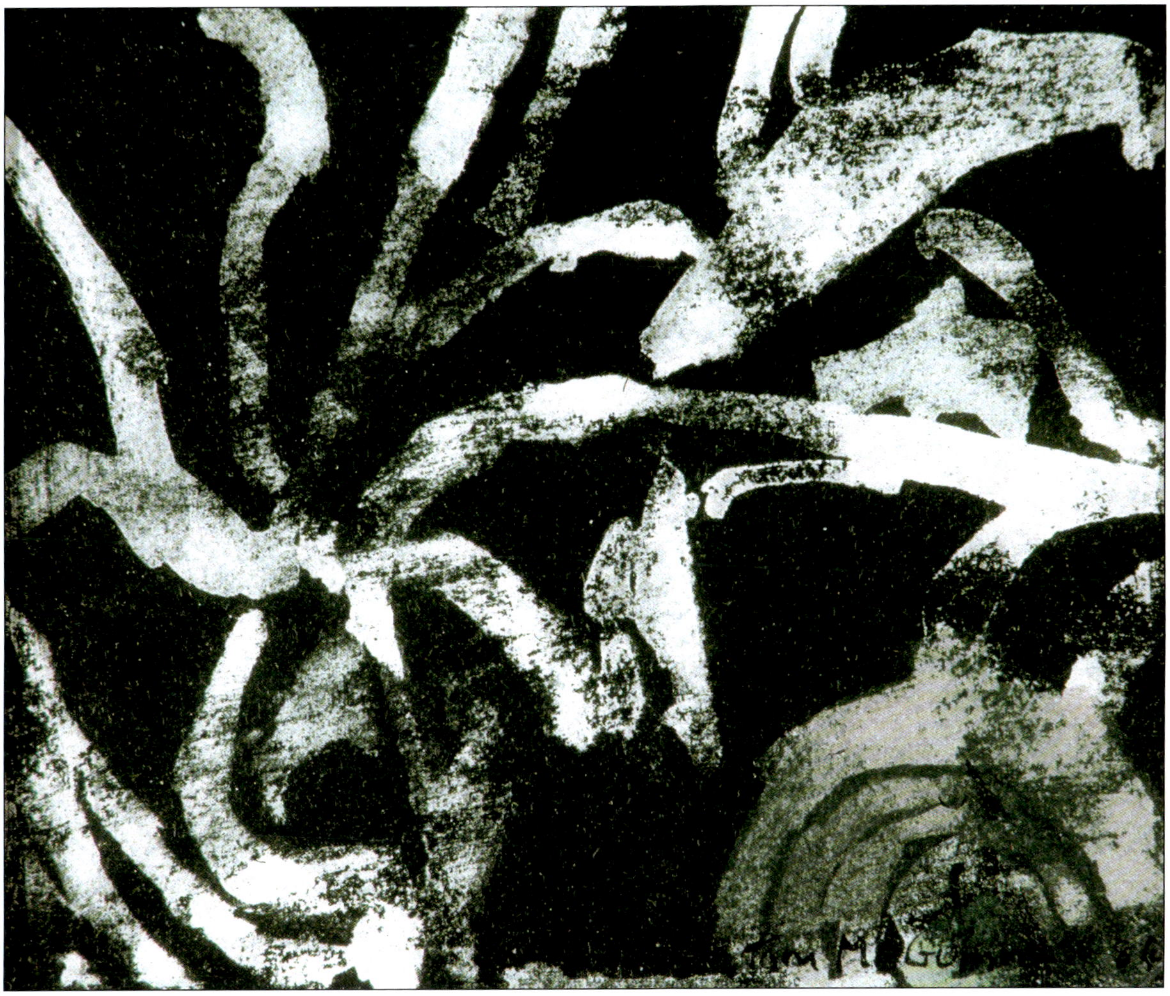

Plate 34

The Decline 1986

Etching - 12 x 10cm

Set against the recent protracted miners' strike of 1984 - 85 this disturbing etching provided "... a chillingly prophetic glimpse into the threats lying over the pits today." Within seven years of this image being made the last deep mine in County Durham closed. The spectre of death seen hovering over the abandoned colliery foretells the demise of the entire coalfield.

See also page 22

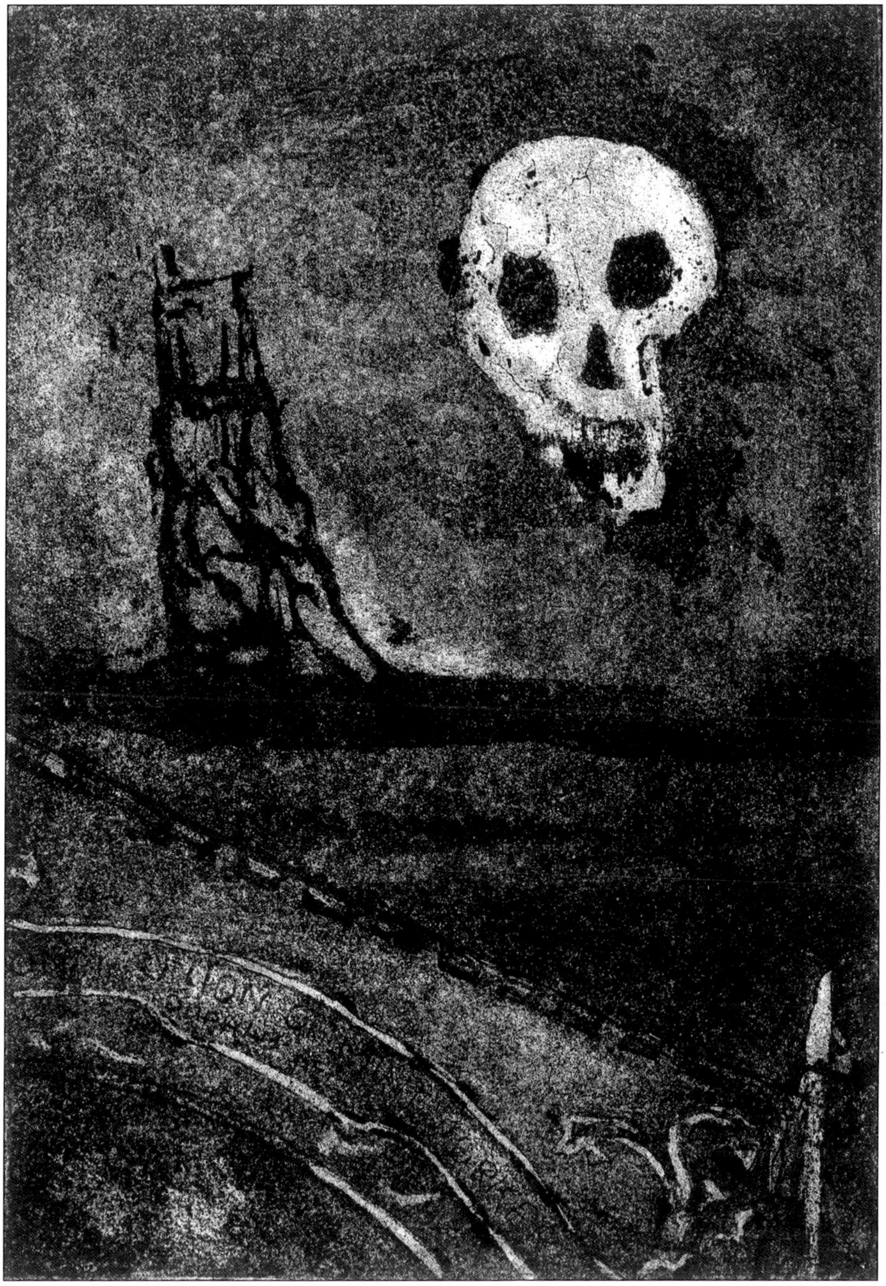

Plate 35

Riding the Skip Winder 1987

Etching - 45 x 55cm

In this large etching Tom introduces us to the vertiginous heights and plummeting depths of the shaft. In a composition that is reminiscent of M. C. Escher's continuous stairway, the eye is compulsively drawn towards the bottom of the shaft, paradoxically seen both as a black abyss and as the provider of economic stabilty for the miner.

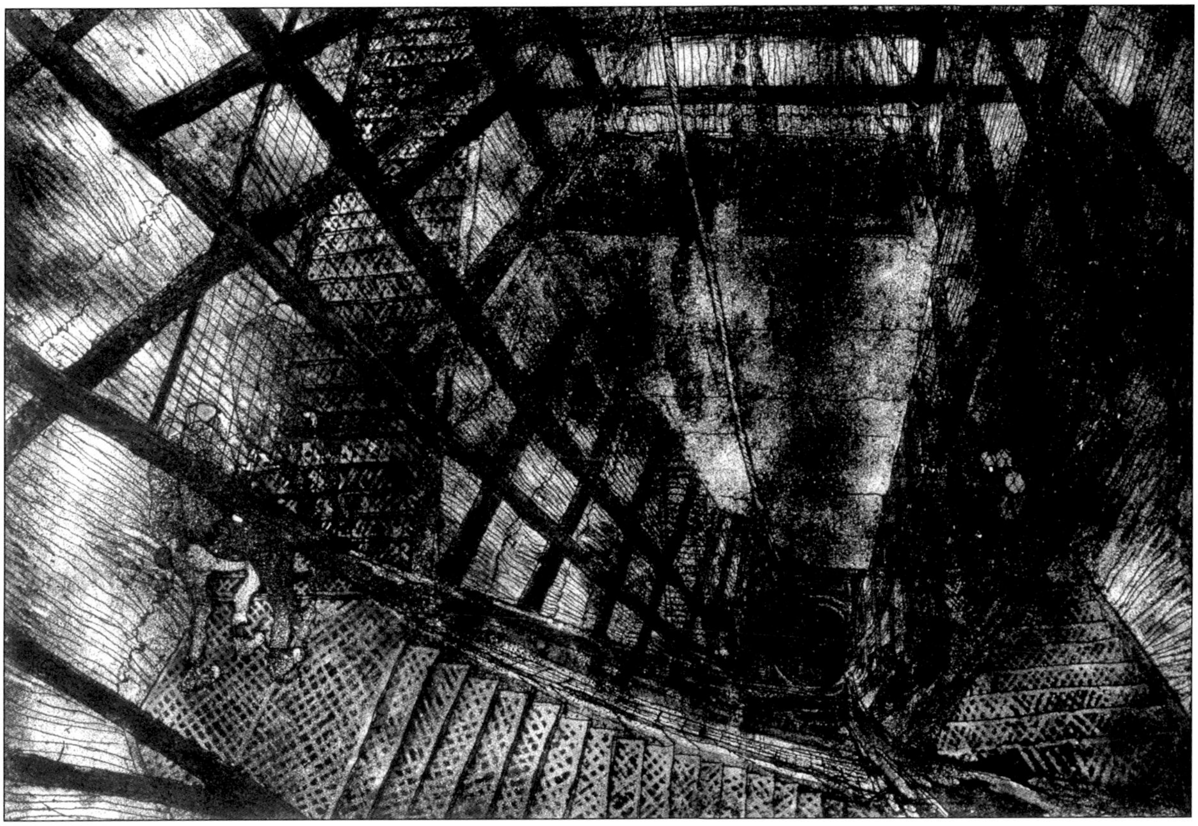

Plate 36

The 'A' Team 1990

Aquatint etching - 10 x 14cm

There is a political statement in this small etching. The 'A' team, a title which Tom coined to denote a team of 'crack men', are about to enter the left hand tunnel with its old but labour-intensive working practices. They are seemingly ignoring the modern, mechanised tunnel on the right. Approaching the men in the distance are the deputies. They will place the men. The dilemma between efficiency and humanity will be decided by others - by the politicians. In the end economics and efficiency would triumph over loyalty and humanity and in this image Tom was foreseeing the final great round of pit closures which would soon end coal mining in County Durham.

See also page 25

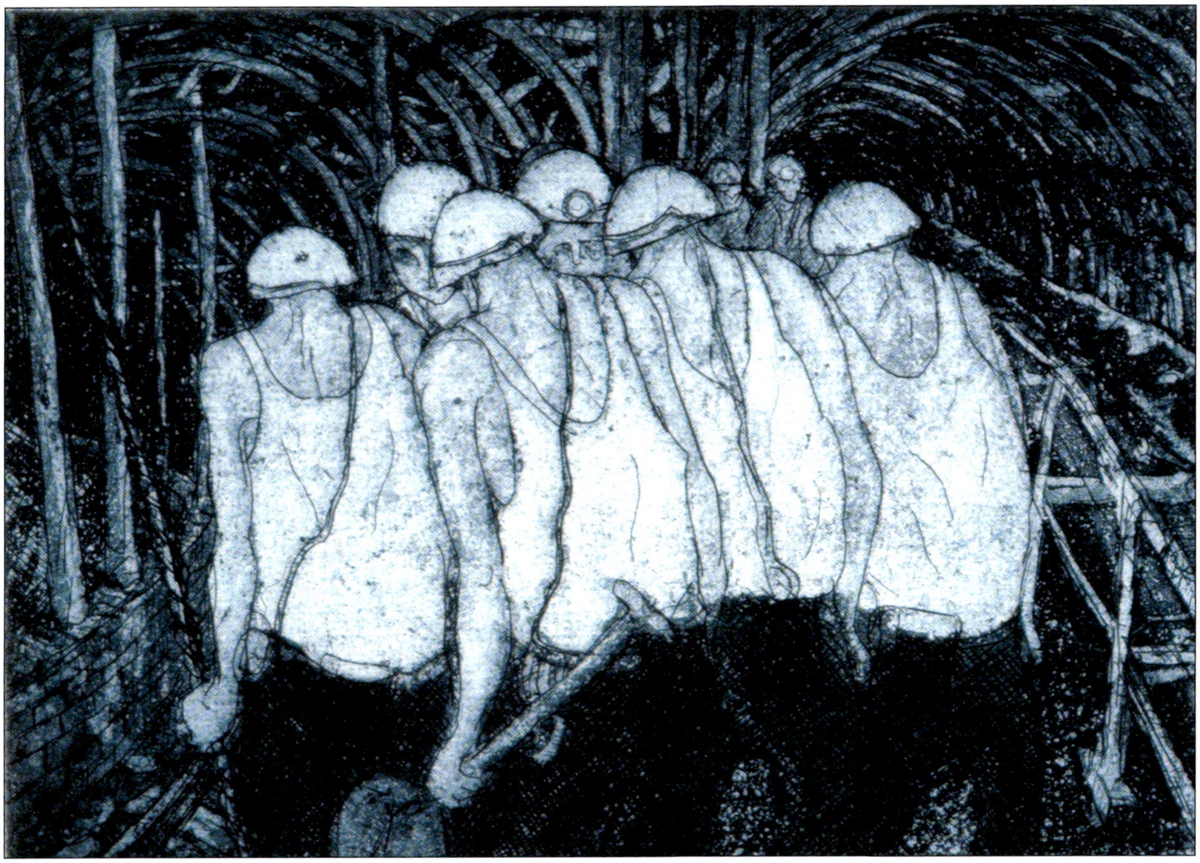

Plate 37

Women Waiting for News 1992

Coloured lithograph - 50 x 35cm

Women waiting for news of their menfolk after a disaster is a recurring theme in art and is a theme to which Tom repeatedly returns. In a scene with resonances of the crucifixion the young women wait outside the colliery gate. The insubstantial figures of messengers approach to inform the women of their men's possible entombment.

See also page 24

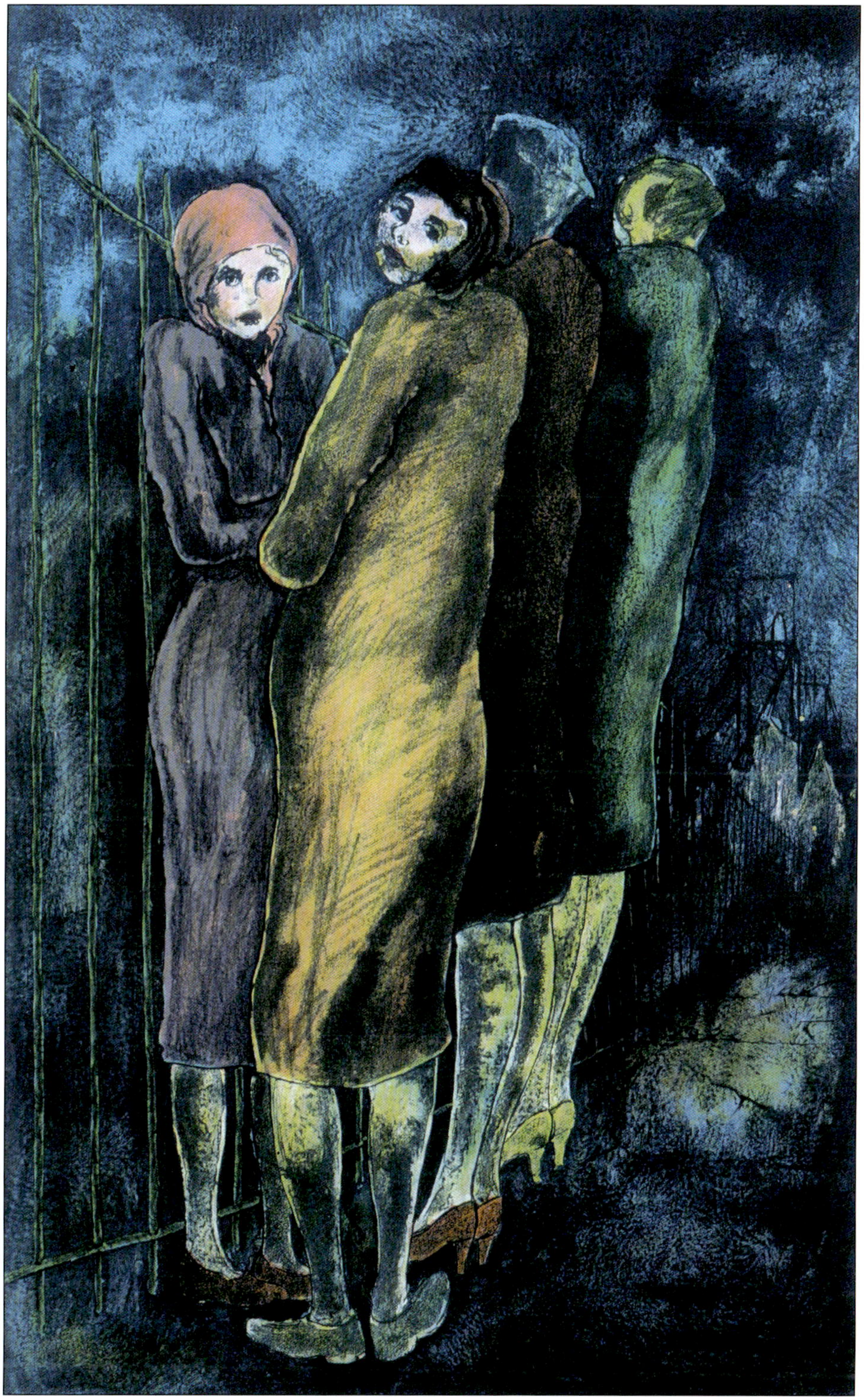

Plate 38

Dark Moon 1993

Sugar lift etching - 15 x 21cm

In a style reminiscent of German artist Otto Dix, this etching is part of Tom's 'Lost Generation Series'. In this series he depicts the now jobless miners as wandering souls, unable to find a new focus in life, continually being drawn back by the memory of the colliery and the nostalgia for the pityard. The men appear as cadaverous, wretched, tormented beings hoping for salvation. In this distressing image the miners are captive victims almost akin to those in a concentration camp.

See also pages 21 & 22

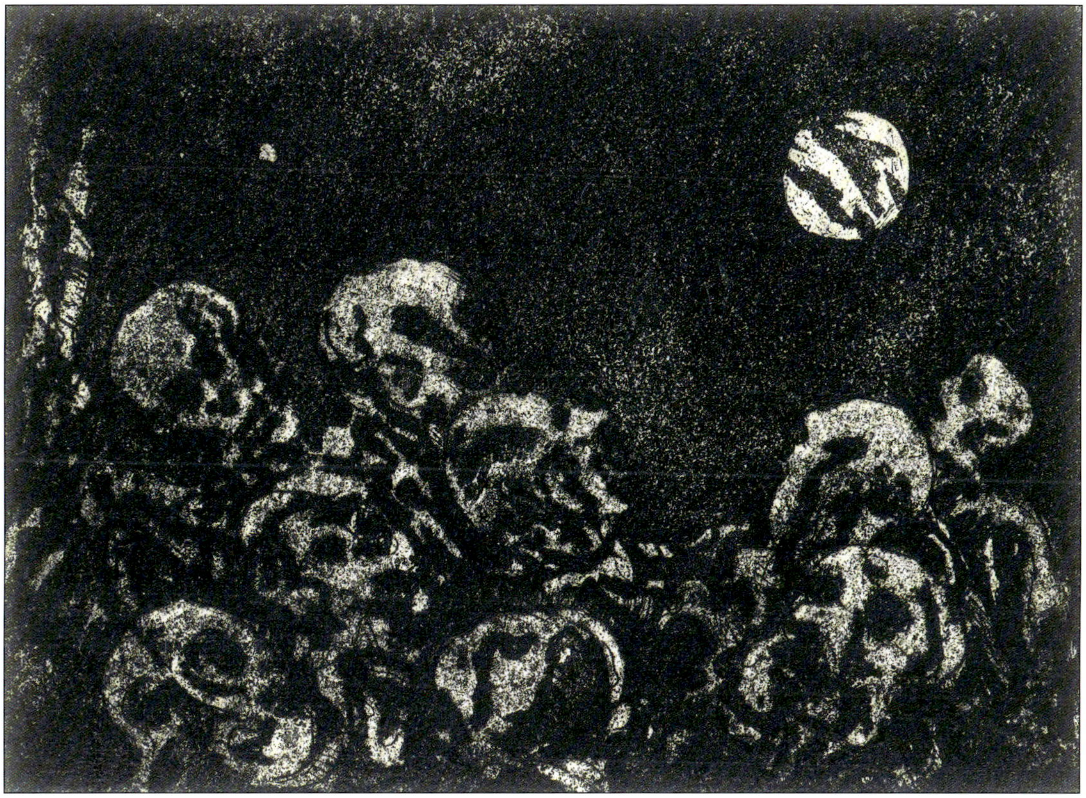

Plate 39

The History of Mining 1989

Mixed media - 105 x 32cm

In this complex composition Tom explores the different phases in mining history by taking a cross-section through the rock to reveal the various coal seams, each worked by an individual from a different era. In a reversal of the archeological norm the most recent is at the deepest level - a level which would only become accessible with modern technology. Through this simple but effective device Tom has created a pictorial coalmining time-line - a technique he also exploited in his 1981 etching ‘2 a.m.’ (Plate 27).

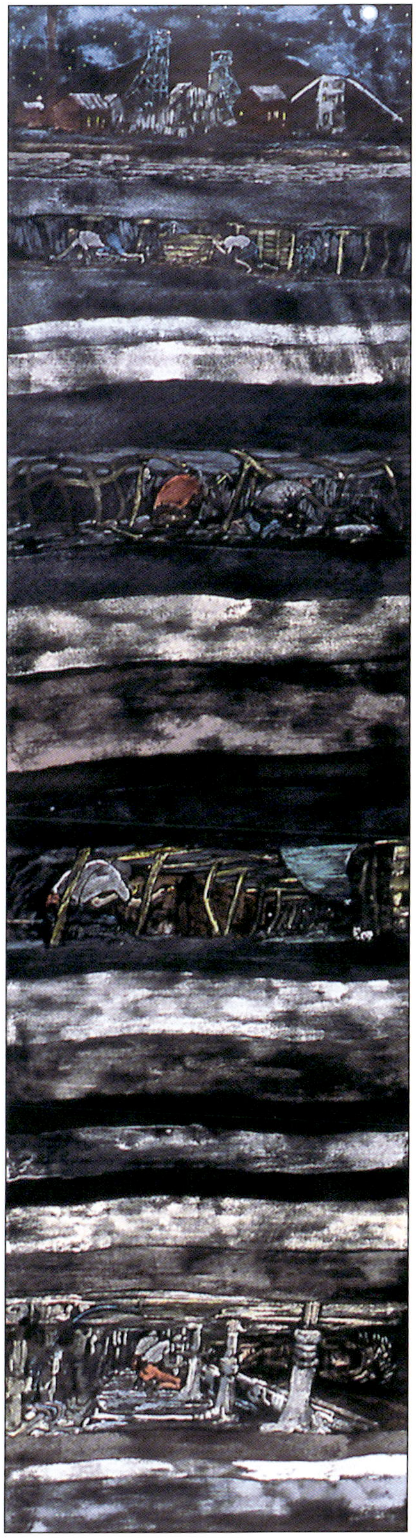

Plate 40

Old Men on the Heap 1996

Oil - 60 x 90cm

The very title of this picture tells us of the desperate plight of these unemployed miners. Tom succeeds in making the viewer feel as uncomfortable as the miners themselves - victims of redundancy - who stare out, begging us to plead their case. The black waste heap and the disused headgear no longer offer any guardianship beneath a sky that appears aflame with indignation.

See also page 22

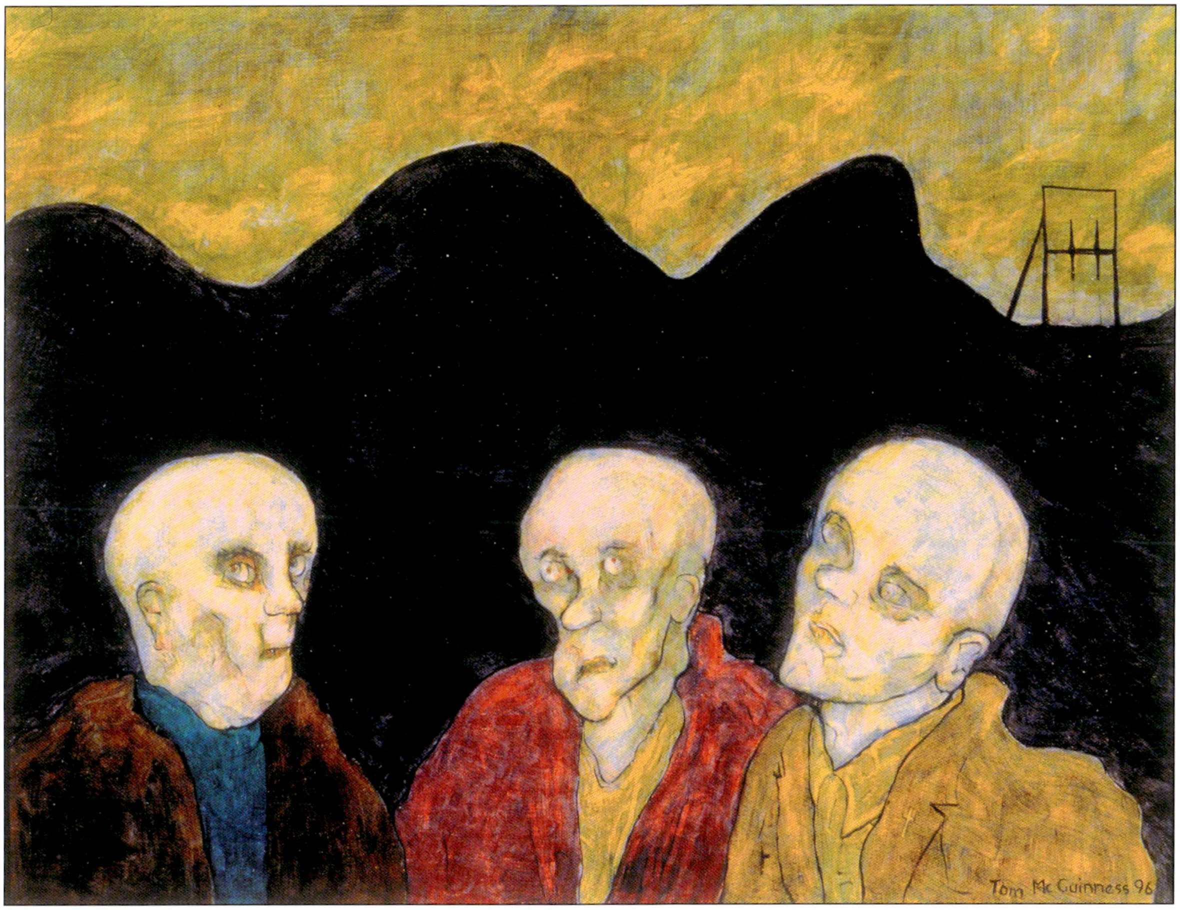
Tom McGuinness 96

Plate 41

The Driller 1994

Mixed media - 55 x 60cm

Exhibited for the first time in 1994 in the Mall Gallery, Crook, this picture demonstrates Tom's wash-off technique to fine effect. Here Tom has given added strength to the contrasting tones and colours by over-painting the periphery of the washed off image with indian ink and by picking out the details of the miner and his drilling equipment in gouache. The effect is most dramatic in juxtaposing the jet black ink against the white highlights of the gouache.

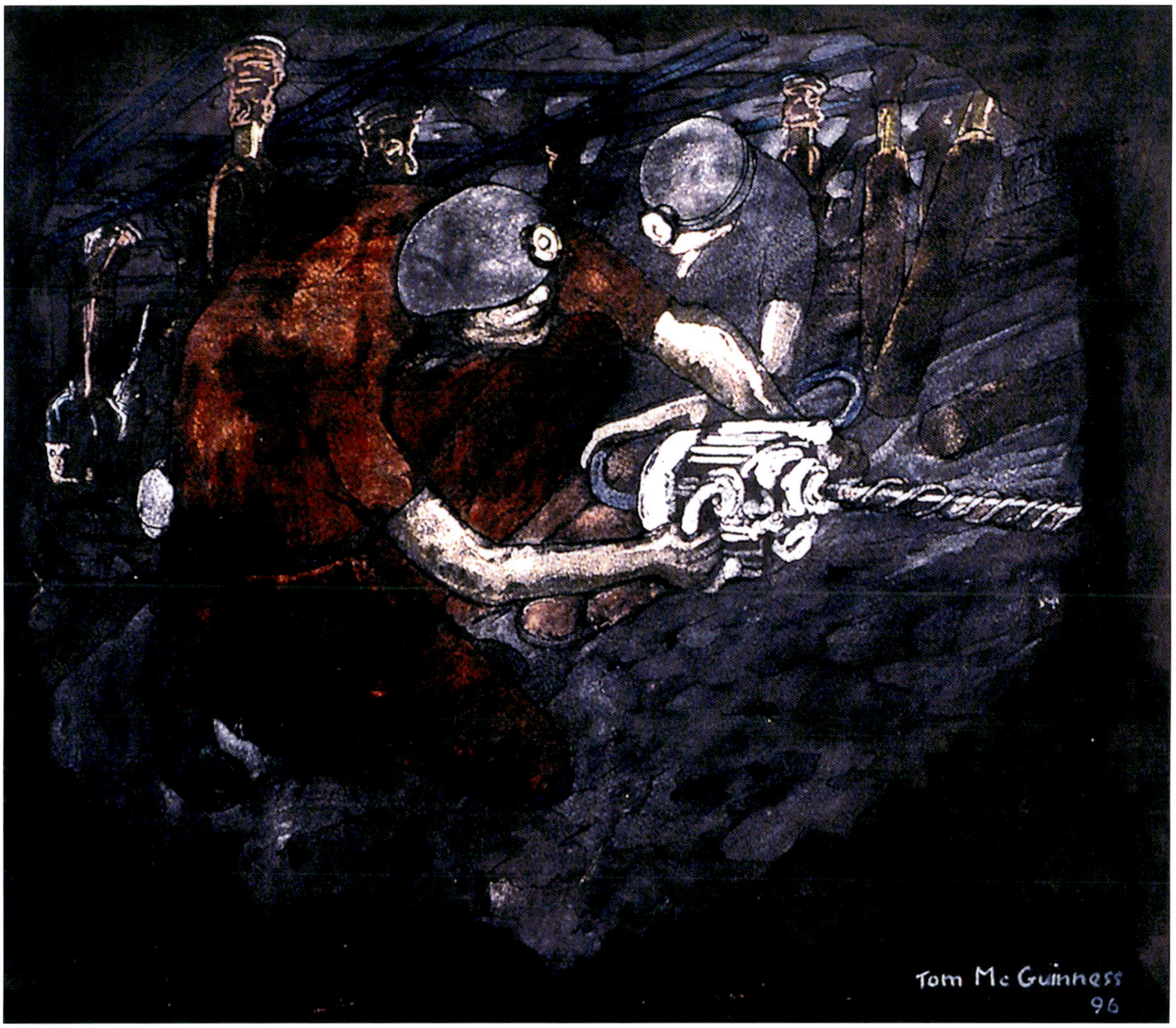
Tom McGuinness
96

Plate 42

Coal Hewer 1996

Oil, canvas on board - 60 x 48cm

This intriguing small glazed oil is not a comfortable painting. The unreal nature of the image, achieved by using glazes in the subtlest shades of veridian and ultramarine, creates a disturbing and lingering impression. The hand hewer tries to bring down loose roof stone, blanched by fear, as he cowers in the left half of the picture. A representative of an industry that is now history, does his ghostly form symbolise the death of a once proud and seemingly indestructible way of life?

See also page 26

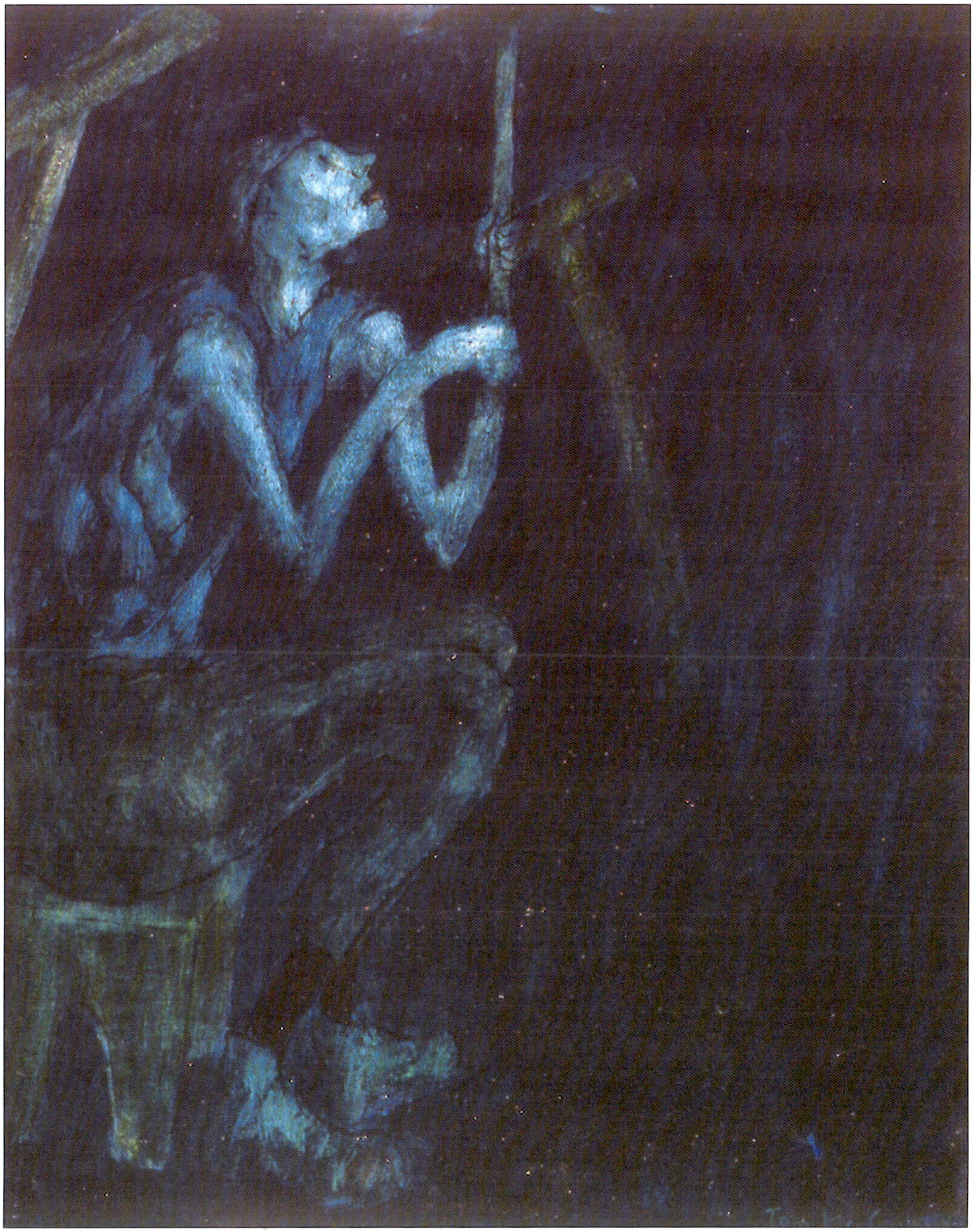

Plate 43

Nightshift 1990

Glazed oil and tempera on canvas - 90 x 30cm

There is no specific location for this picture, but Nightshift represents a recurring theme in Tom's work and there are similar images to be found in his etchings and in a stained glass window in the McGuinness Gallery at Bishop Auckland Town Hall. Yet again, the men trudge compulsively to repeat their ritualistic task, drawn inexorably towards their fate. The use of tempera to highlight the fine detail is accentuated by the moon's soft glow which the glazes confer.

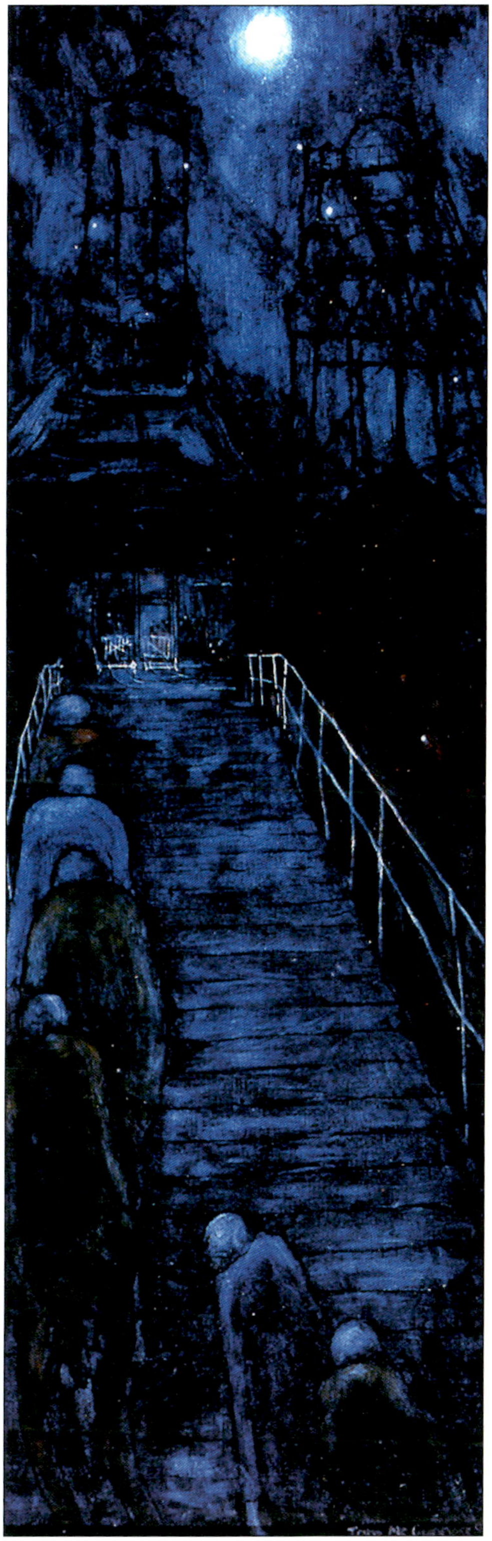

Plate 44

Harvey Wetmen 1996

Glazed oil, canvas on board - 68 x 90cm

This image was used for the cover of the biography, 'Tom McGuinness - The Art of an Underground Miner', Gemini Productions 1997. Compared with his recent Lost Generation series this painting represents Tom's return to the security of a working colliery. The figures turn away from the observer, unlike the Lost Generation they no longer need to seek the support that eye contact with the viewer confers, as their glistening, vivid yellow waterproof jackets contrast with the bright moonlit sky. The colours have become more vibrant - a development that Tom was to continue.

See also page 23

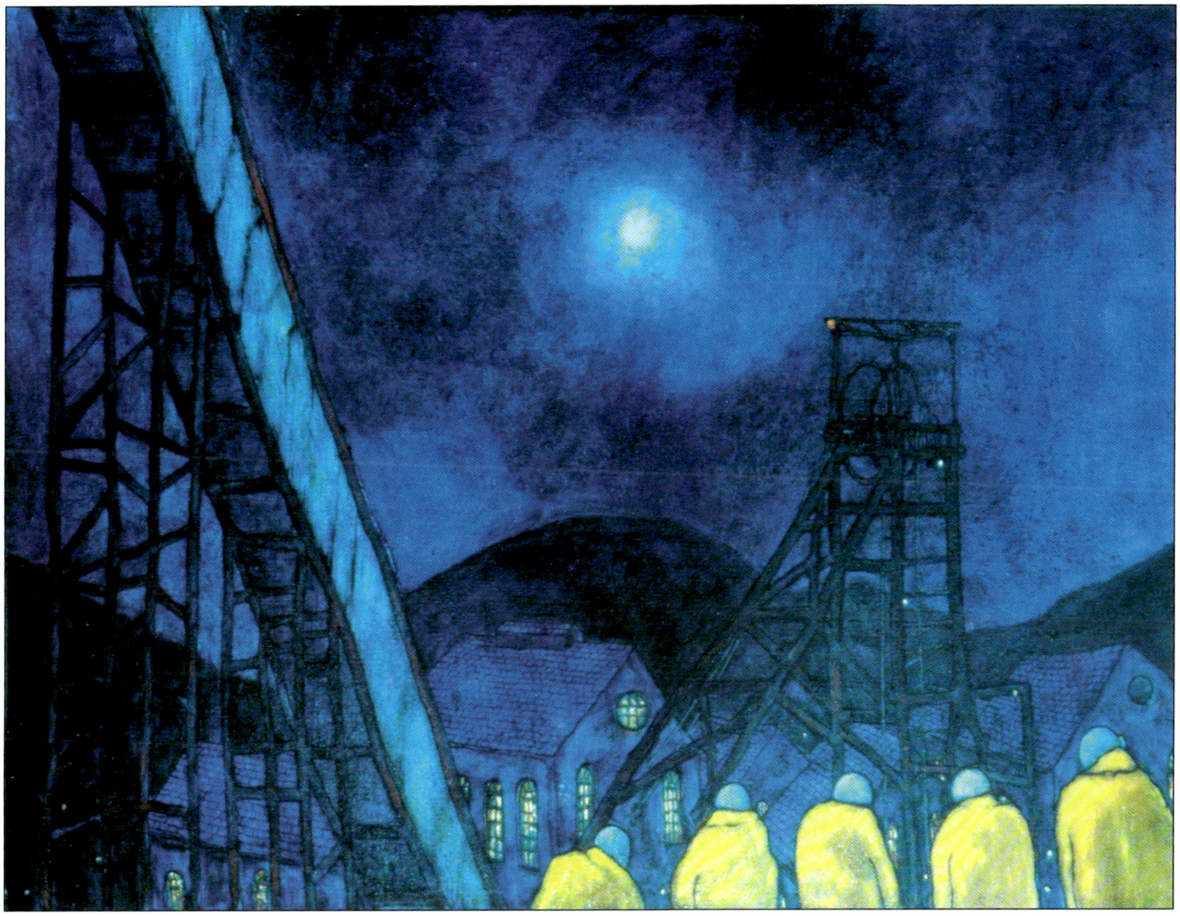

Plate 45

East Hetton 1996

Oil on canvas - 90 x 120cm

All the drawing for this work were done on site at East Hetton Colliery in 1982. The final image is an amalgam of many drawings, each of a detail within the picture. Tom has produced two major versions - the first in 1982 using his wash-off technique which he donated to Cassop cum Quarrington School to enhance their permanent mining exhibition. Tom found this to be one of his most satisfying pictures to paint, despite the asymmetry of his hand-made canvas which made subsequent framing impossible.

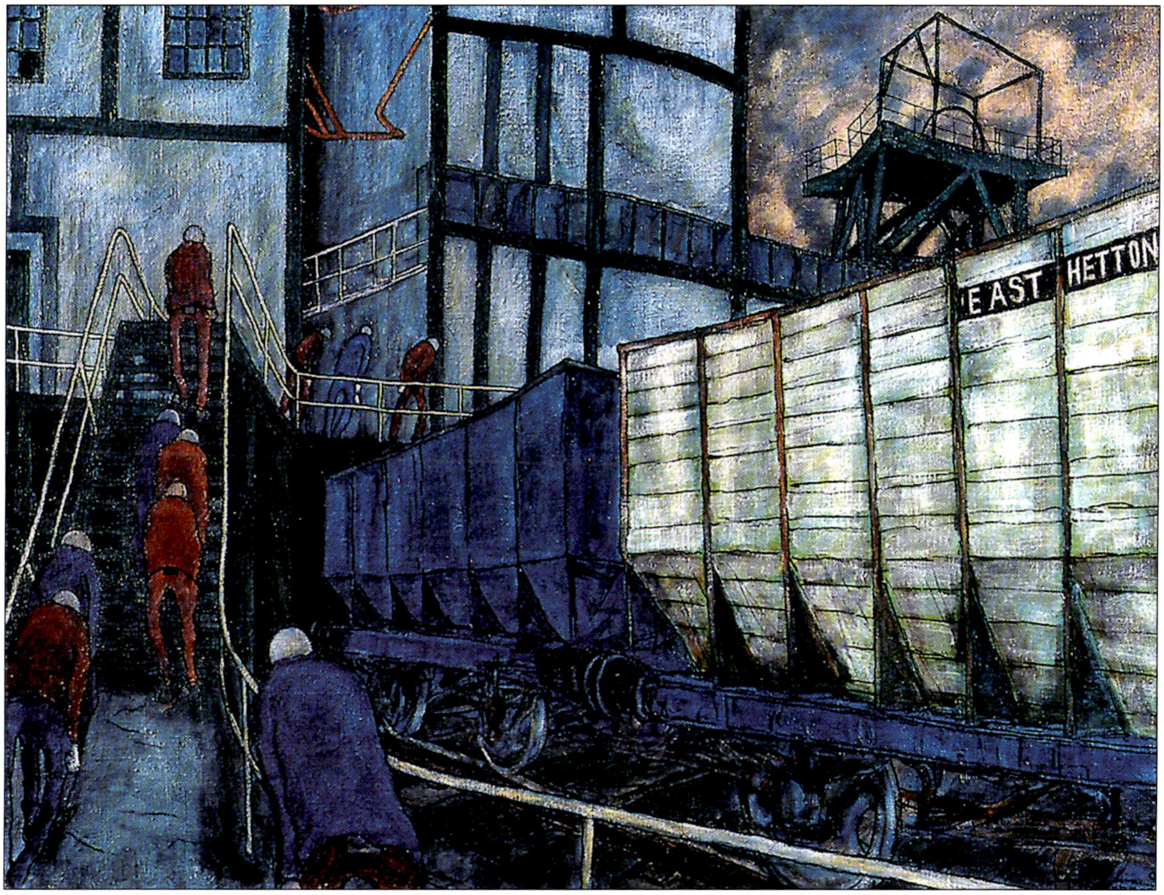
EAST HETTON

Plate 46

Men Riding the Skipwinder 1987

Mixed media - 100 x 45cm

In this virtuoso performance of the craft of mixed media Tom has converted the skeletal structure of the mine shaft into a futuristic design - a design distilled to a pattern. From a distance there is an amazing sense of depth and on closer inspection the tiny men reinforce the colossal scale of the undertaking. Tom has produced two versions of this theme - the other in oil - but it is because of the imprecise nature of the wash-off technique that this version stands out as one of Tom’s most imaginative and accomplished works.

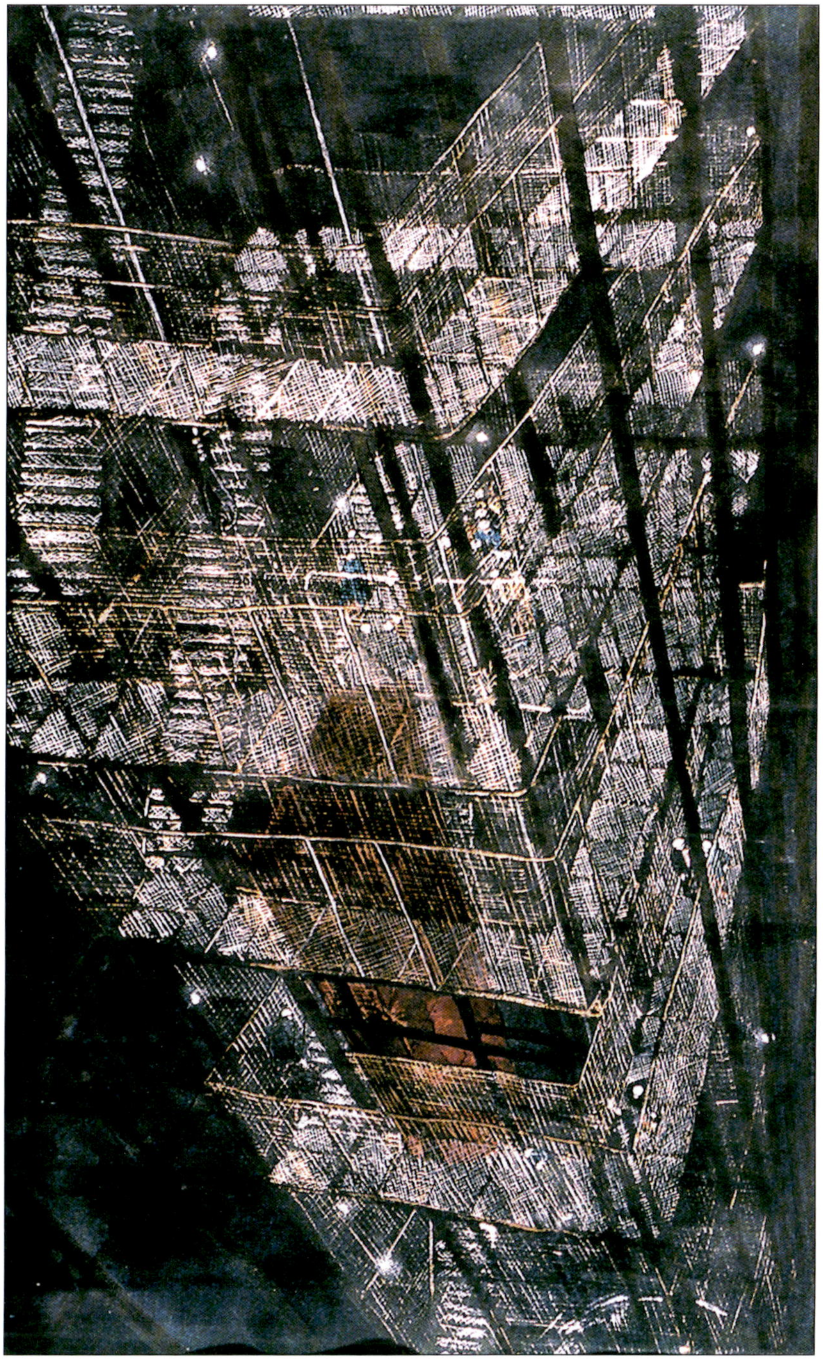

Plate 47

Miners Going to the Shaft 1996

Glazed oil and tempera on canvas - 120 x 100cm

A perfect example of how intense vibrancy can be achieved from a limited palette. This painting glows in the light of a McGuinness moon which illuminates the minimalised headgear. This has been reduced to the barest outline to give strength to the image of men trudging to work. Here Tom has used tempera to get the fine crisp line of the winding cable.

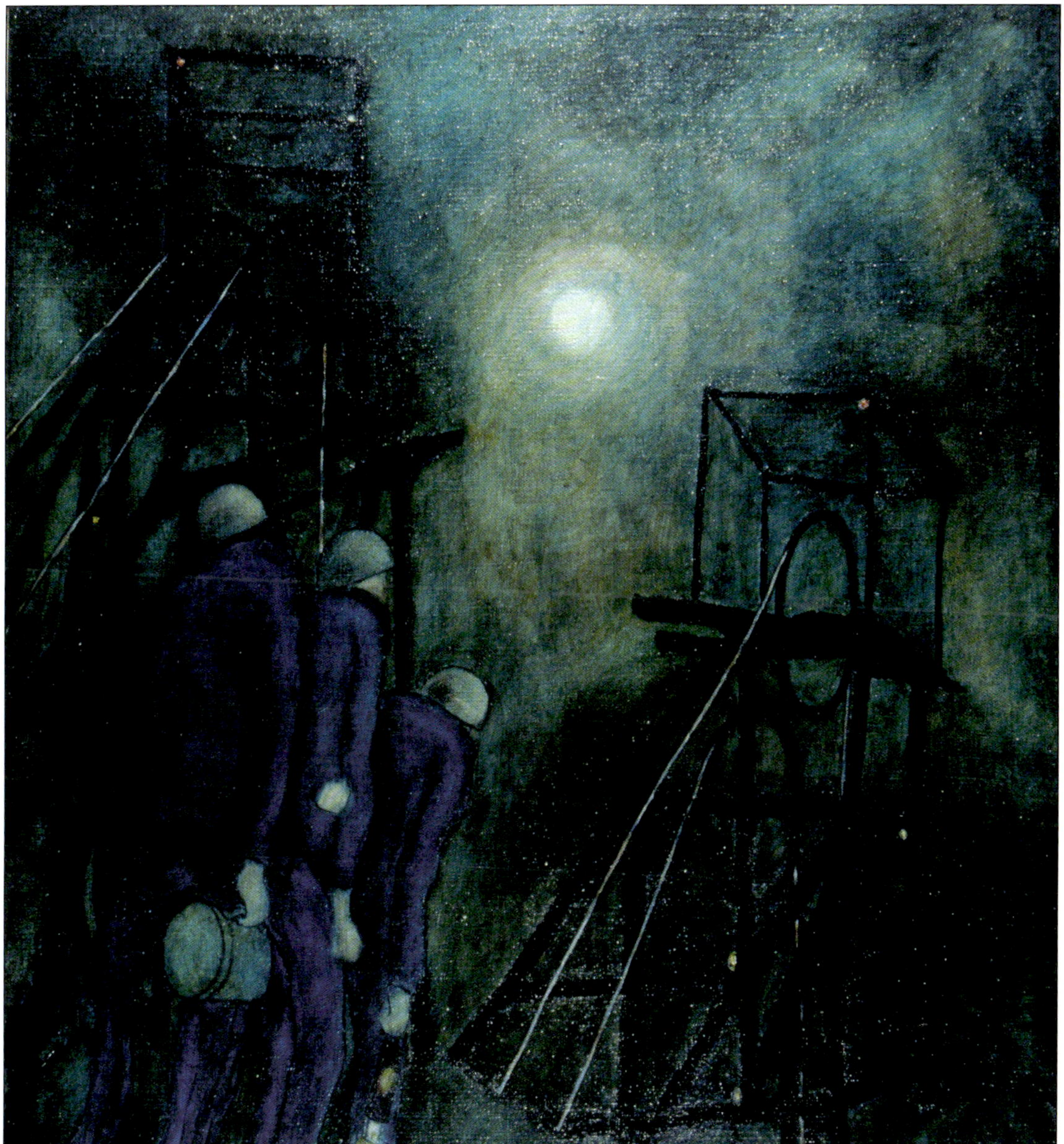

Plate 48

Snow Scene 1996 (unfinished)

Oil canvas on board - 70 x 110cm

This is an imaginery scene of a colliery village in winter - a theme that Tom has frequently returned to, using a variety of media. He described this as 'a scene after Breughel'. Interestingly Tom has used the right hand side of this image as a personal Christmas card.

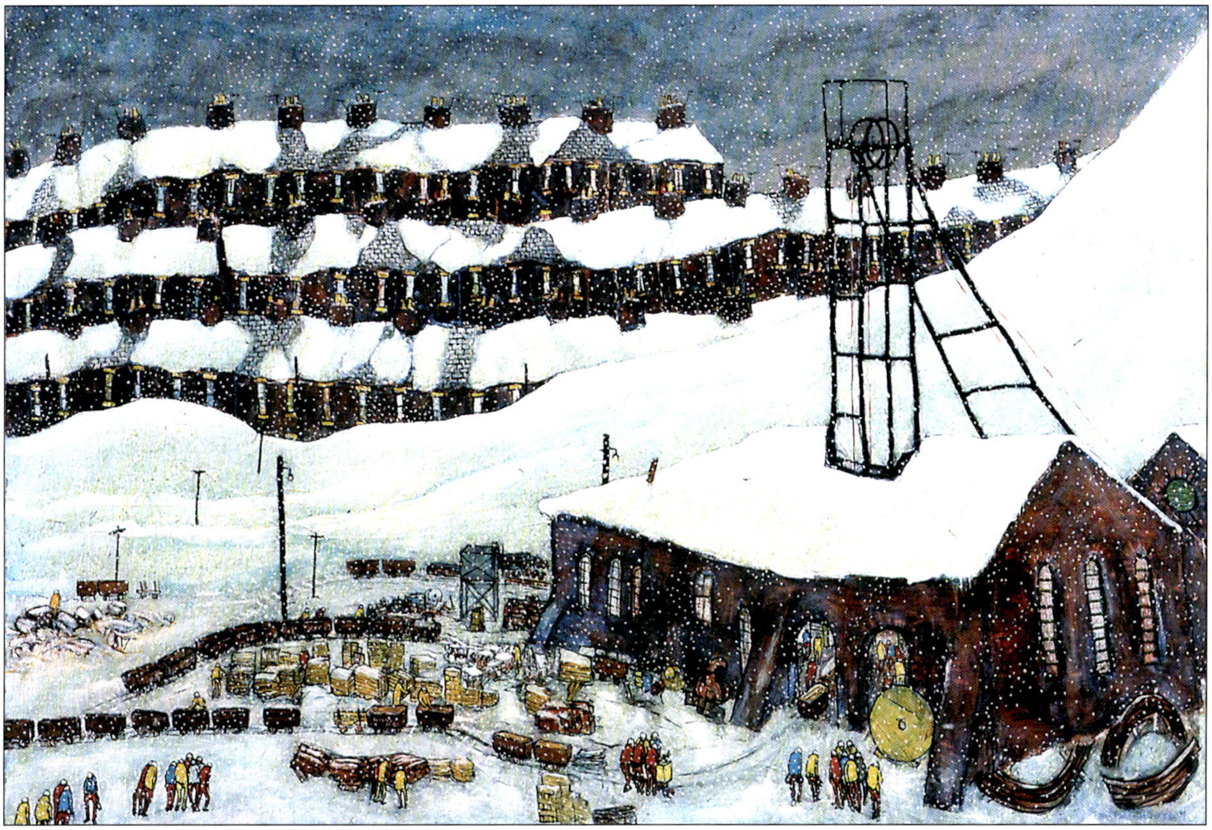

Plate 49

Cages in the Shaft - detail 1997

Stained glass window - 200 x 100cm

Made for the shaft-like stairwell of the Hammersmith home of Doctor Rod Hughes, this commission was Tom's first direct representation of mining life in stained glass.

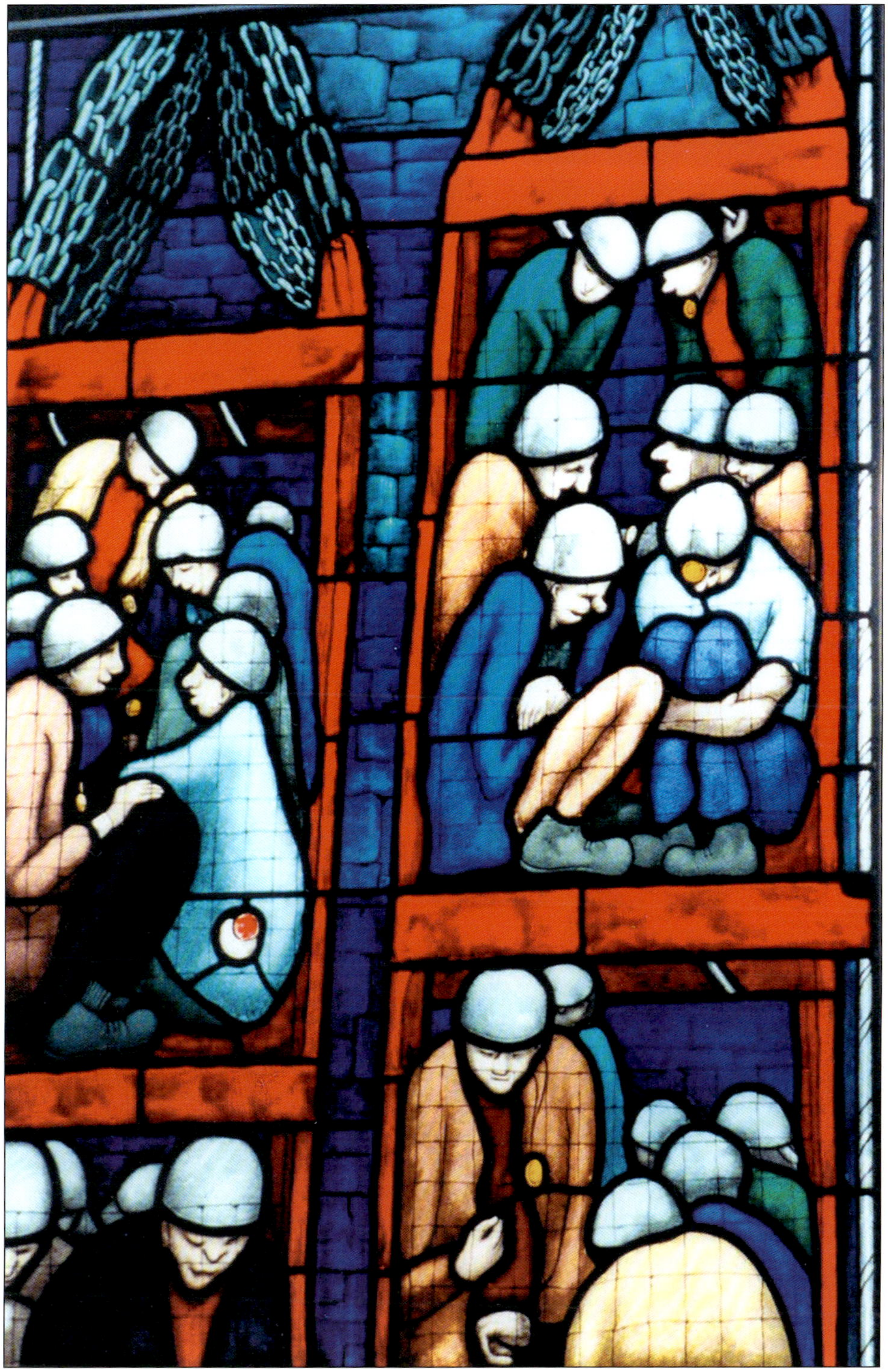

Plate 50

Miners Going to the Shaft 2002

Glazed oil on canvas - 75 x 50cm

This vibrant glazed oil confirms Tom's continuing quest for colour. The picture was exhibited at 'The Shaft' exhibition at Bishop Auckland Town Hall in 2002, which coincided with the naming of the gallery after Tom - The McGuinness Gallery - and his presentation of the stained glass window 'Steps to the Shaft' to the people of Bishop Auckland.

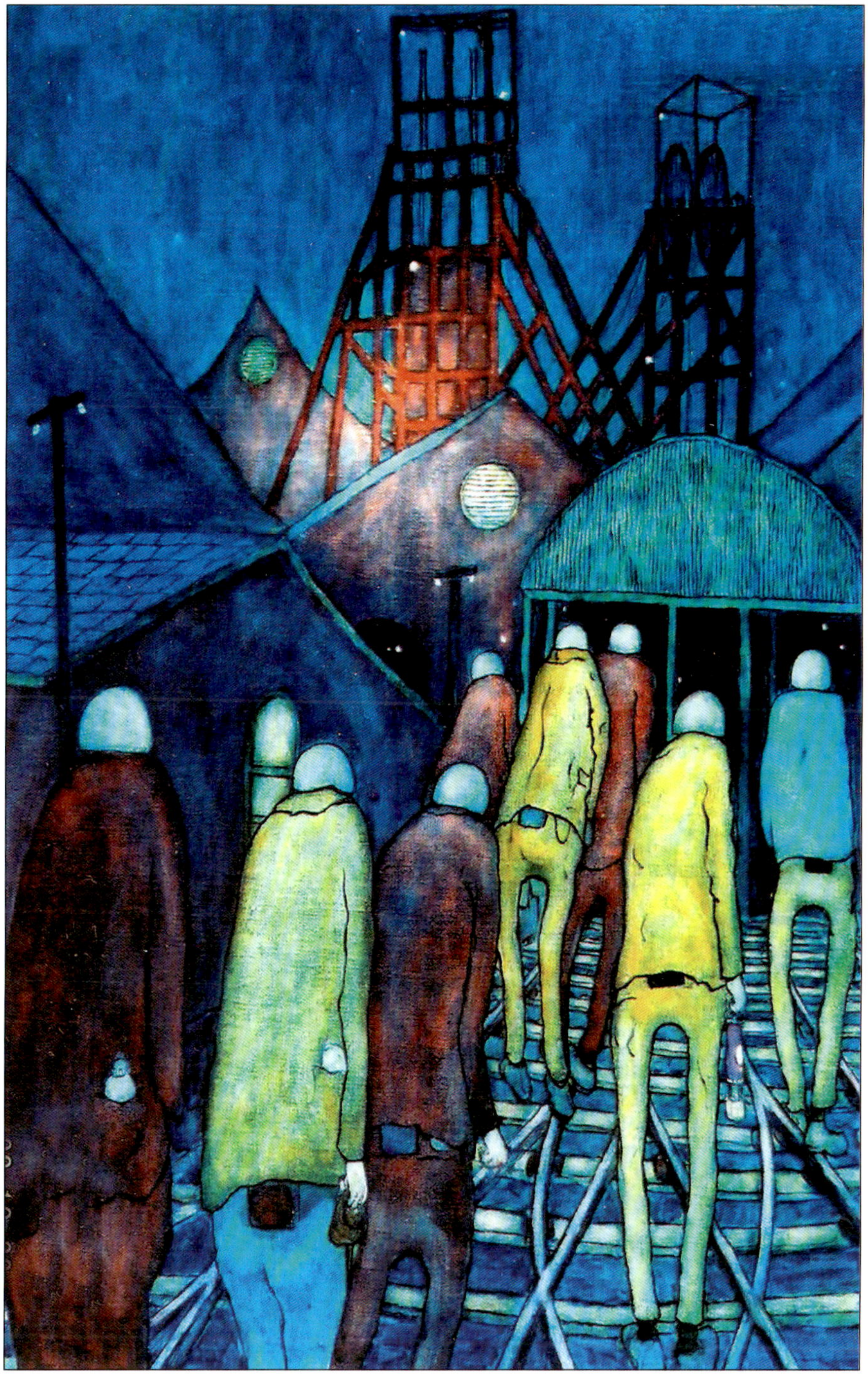

Plate 51

The Pumpmen 2000

Glazed oil on canvas - 130 x 65cm

Set in an almost abstract background, Tom has used the intricacies of the cables, rails, girders and pipes to enmesh the waterproof-clad pump men. Tom had intended to include water in the picture but felt the composition to be complete as it now stands.

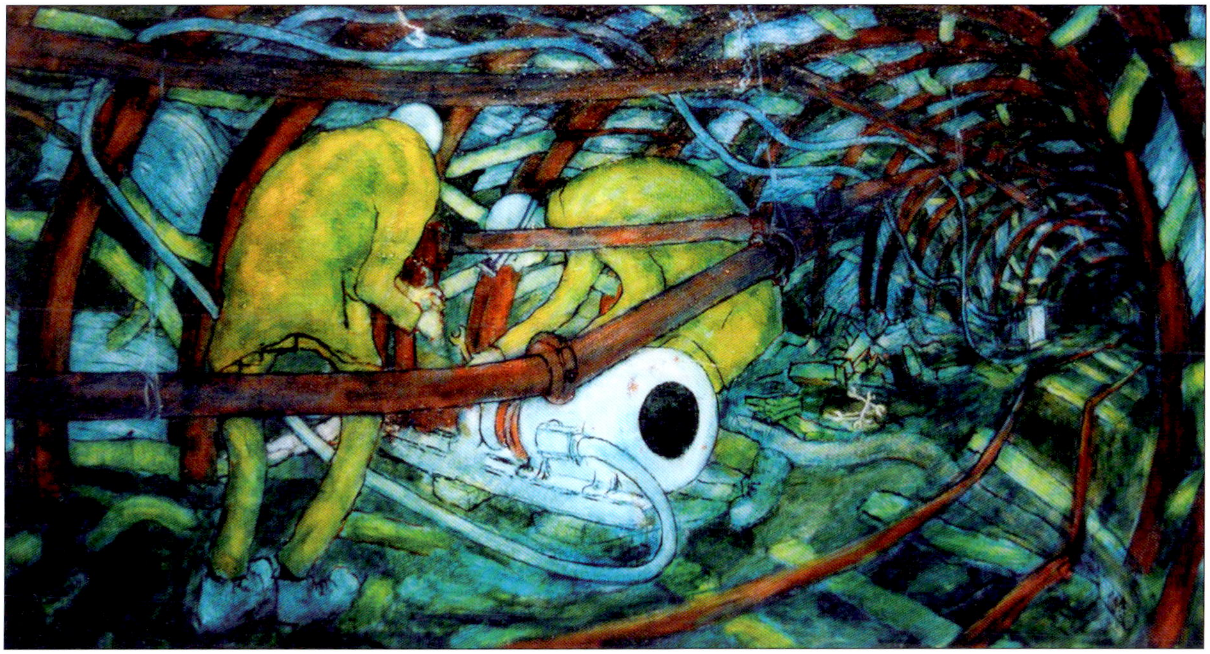

Plate 52

Third Man's Got the Ticket 2001

Glazed oil on canvas - 66 x 74cm

We look curiously down on the men preparing to ride, feeling their imperative, as they dance their way to the cage, urged on by the deputy. They arrive hunched in preparation for their descent. The bright red outfits are in total contrast to the black gaping void into which they are about to be consigned

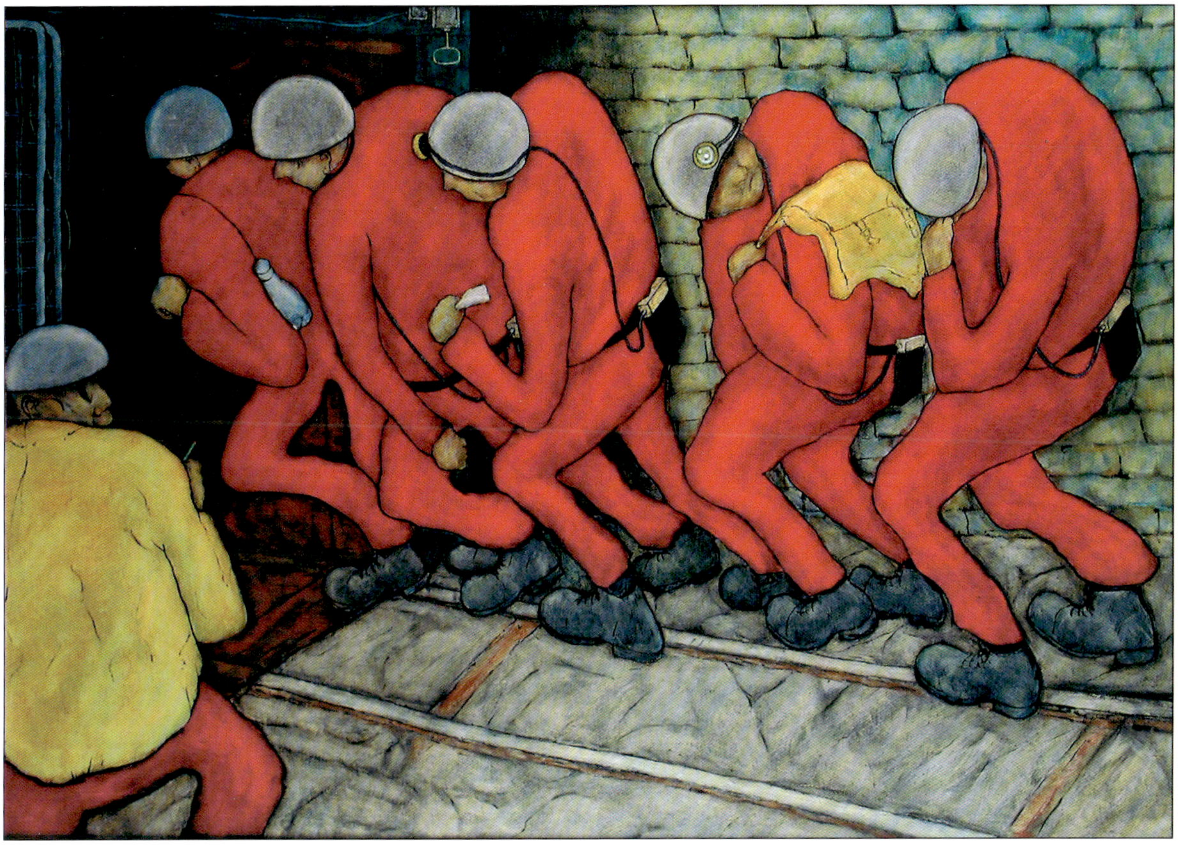

Plate 53

Miners Waiting to Ride 2003

Crayon and graphite - 60 x 80cm

This picture is one of a number of very detailed drawings that Tom has recently executed. He has changed his medium using monochrome crayon and graphite pencil. These studies were done as a form of relaxation in that he was able to sit comfortably in his sitting room with his drawing board on his knee. All have been drawn from memory with reference to his copious sketch-books.

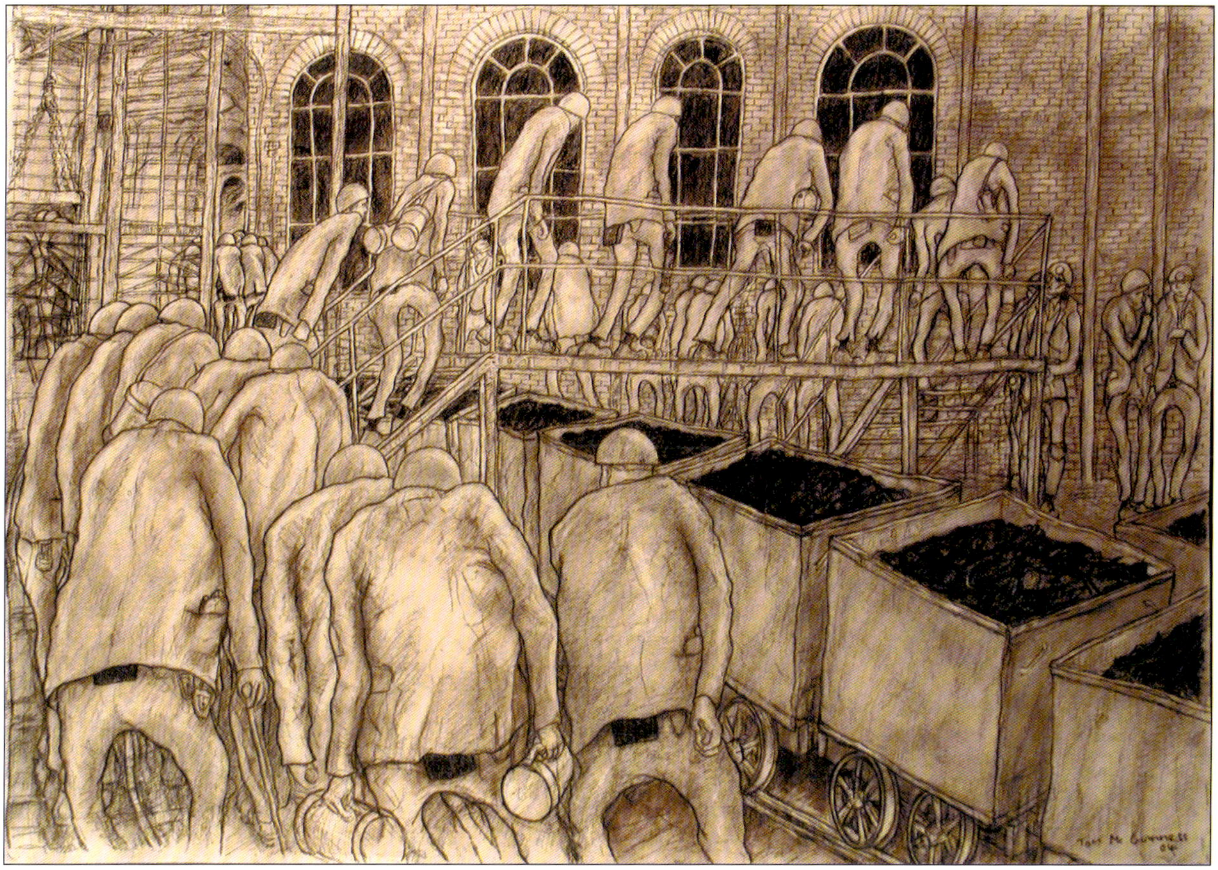

Plate 54

Miners Passing the Deputy 2003

Glazed oil on canvas - 75 x 60cm

In this very colourful picture, first exhibited at Gallerina, Darlington in 2004, Tom has abandoned the eerie blue-green light of his earlier underground scenes in favour of bold primary colours. Unusually the men travelling the returns are facing us and are realistically represented, making the viewer feel comfortable with the image.

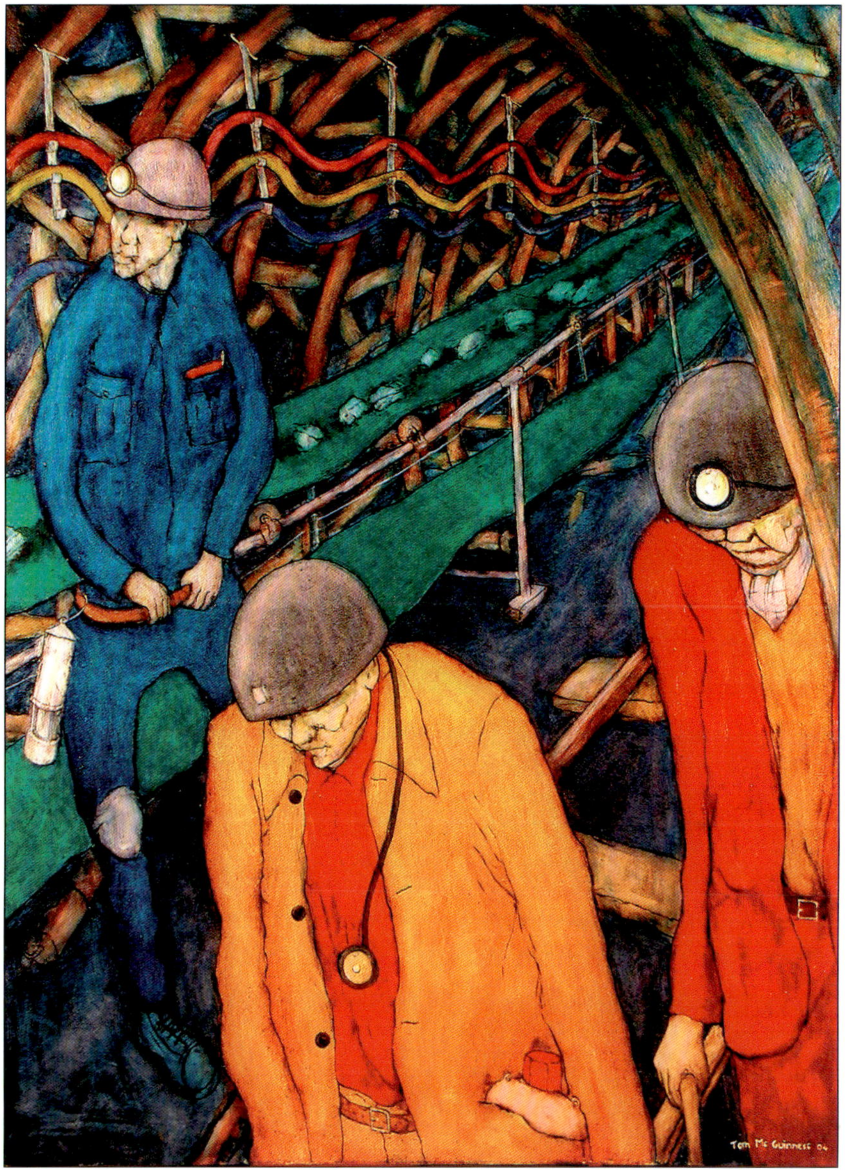
Tom Mc Guinness 04

Plate 55

Miners Passing the Return Wheel 2005

Glazed oil on canvas - 120 x 100cm

Tom has frequently painted this scene. In all of the other versions, however, the miners pass in front of the wheel. Here Tom experiments with design. The wheel in the foreground is shown in stark silhouette against the glowing, glazed Prussian blue tunnel. He has reversed the usual convention and the background becomes the light source. The distorted miners are integral to the overall design.

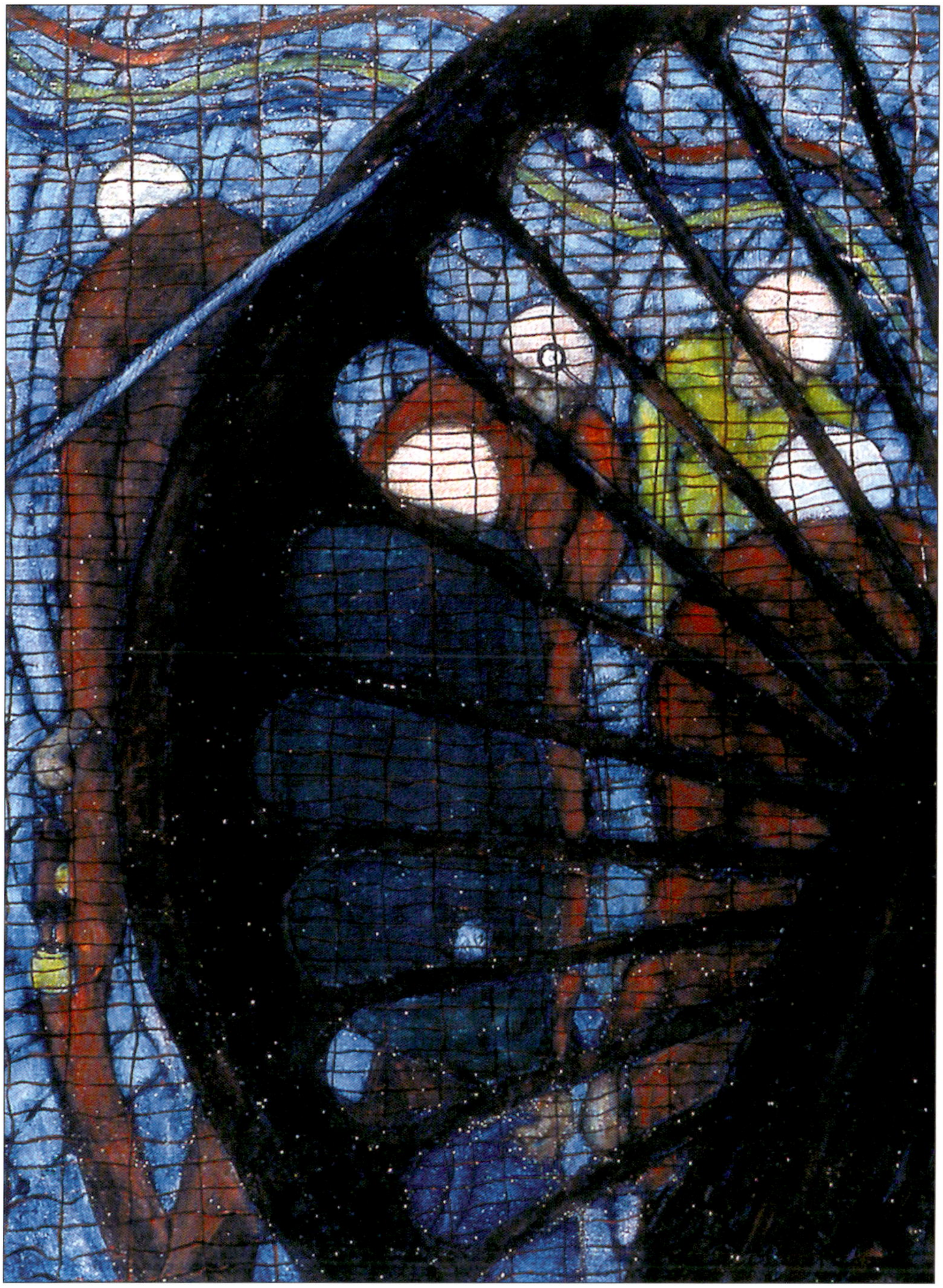

Plate 56

The 'B' Team 2005

Etching - 17 x 8cm

The 'B' Team, a title coined by Tom, is his most recent etching. Interestingly this was the first time he had etched using an aluminium plate. He had read about this technique, fascinated by the dense black line that the medium offered. The image proved difficult to print but must rank as the high point of his etching career. With the assistance of artist and graphic designer, Elaine Kenyon, Tom produced this dry point etching whilst working in the cellar of the McGuinness Gallery at Bishop Auckland Town Hall where his printing press now resides.

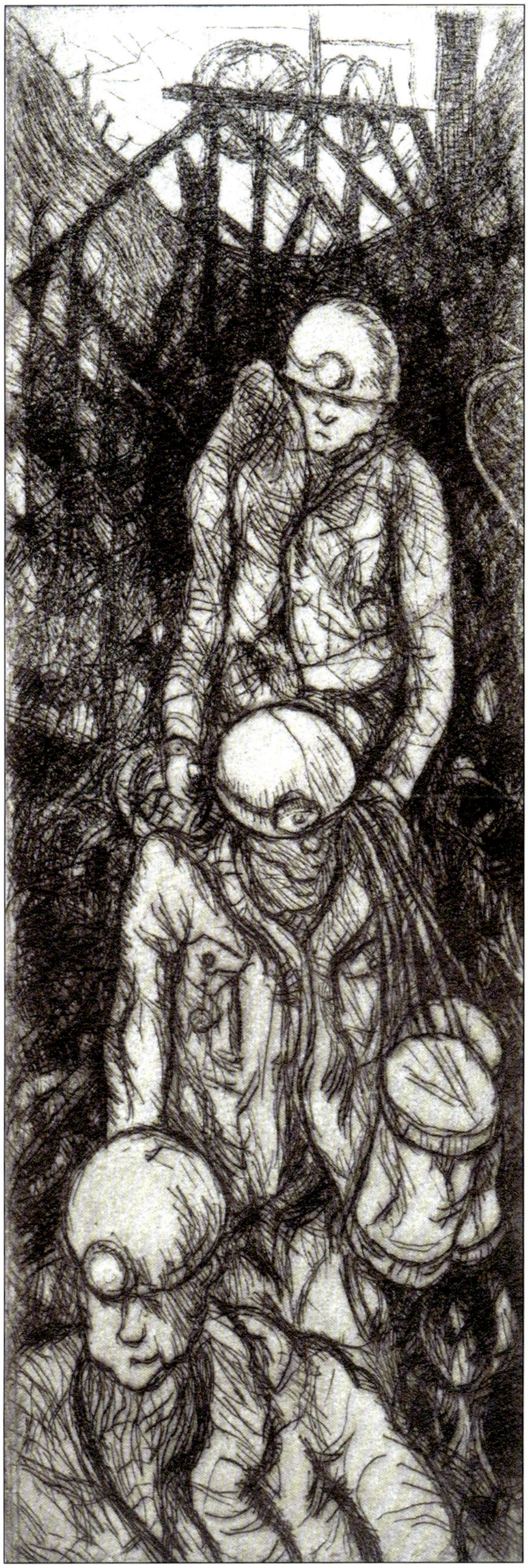

Plate 57

Resting on the Way 2005

Stained glass window - 110 x 70cm

Crafted by Duncan Storr and manufactured at the North Grange Glassworks, Newcastle

Commissioned for St. Joseph's RCVA Primary School, Coundon, (the school attended by Tom's grandsons), as part of a National Lottery funded mining project, Tom's design compassionately places The Holy Family in modern times. Resting on a park bench on the edge of a mining village, Joseph is shown as a miner, complete with Northern Echo and bottle of Brown Ale. Mary, attending the infant Jesus, is resplendent in headsquare and mackintosh whilst a pit pony, replacing the donkey, takes a well-earned rest. Tom places the scene in the 1940s showing a miner who, having been discovered as a union member, has not only lost his job but as a result he and his family are now homeless - a direct analogy to the Holy Family.

See also page 18

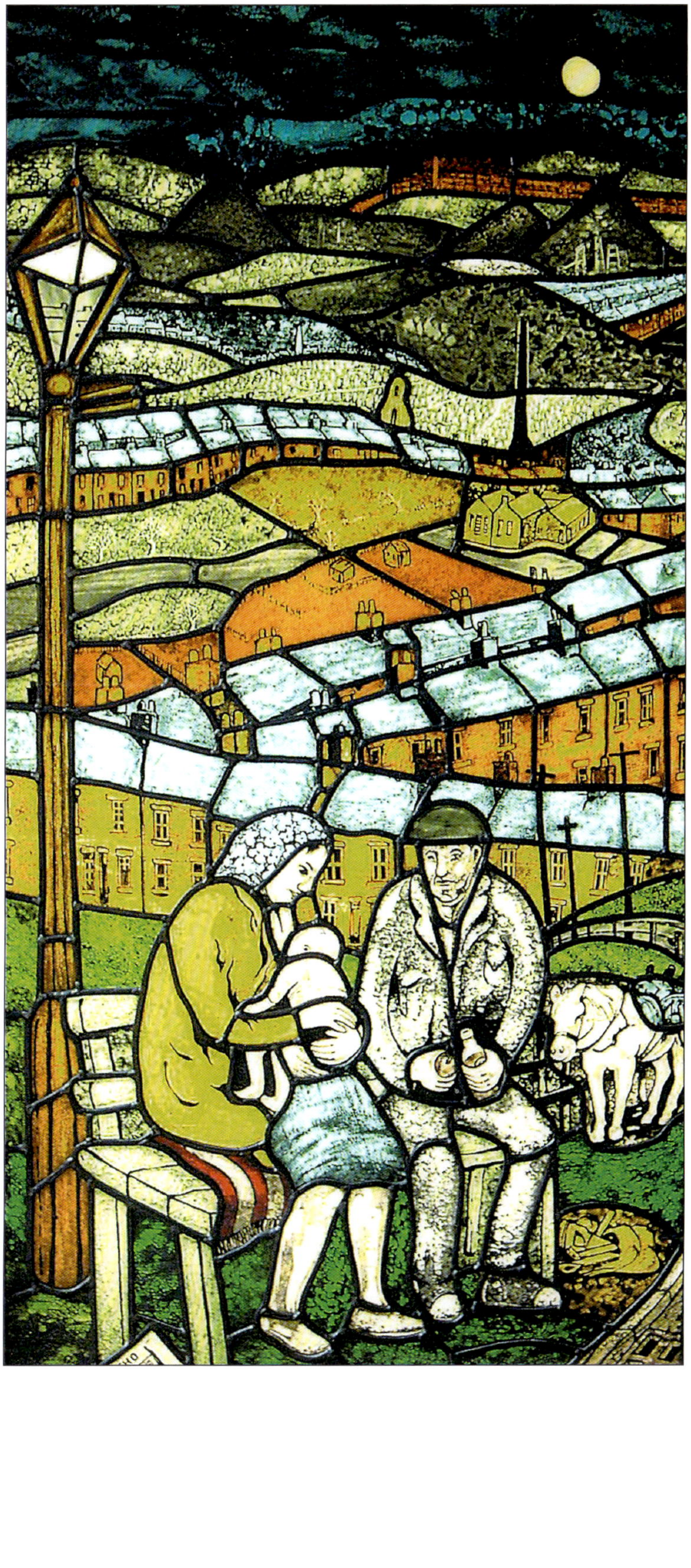

Plate 58

Sketch-book various dates

Mixed media on paper

A page from one of Tom's hundreds of sketch-books. These sketches provide Tom with a huge, first-hand, contemporaneous resource. They constitute an invaluable aide-memoire for the artist and are a fascinating visual record of the minutiae of the miner's world. Tom has used these sketch-books throughout his career to construct major compositions.

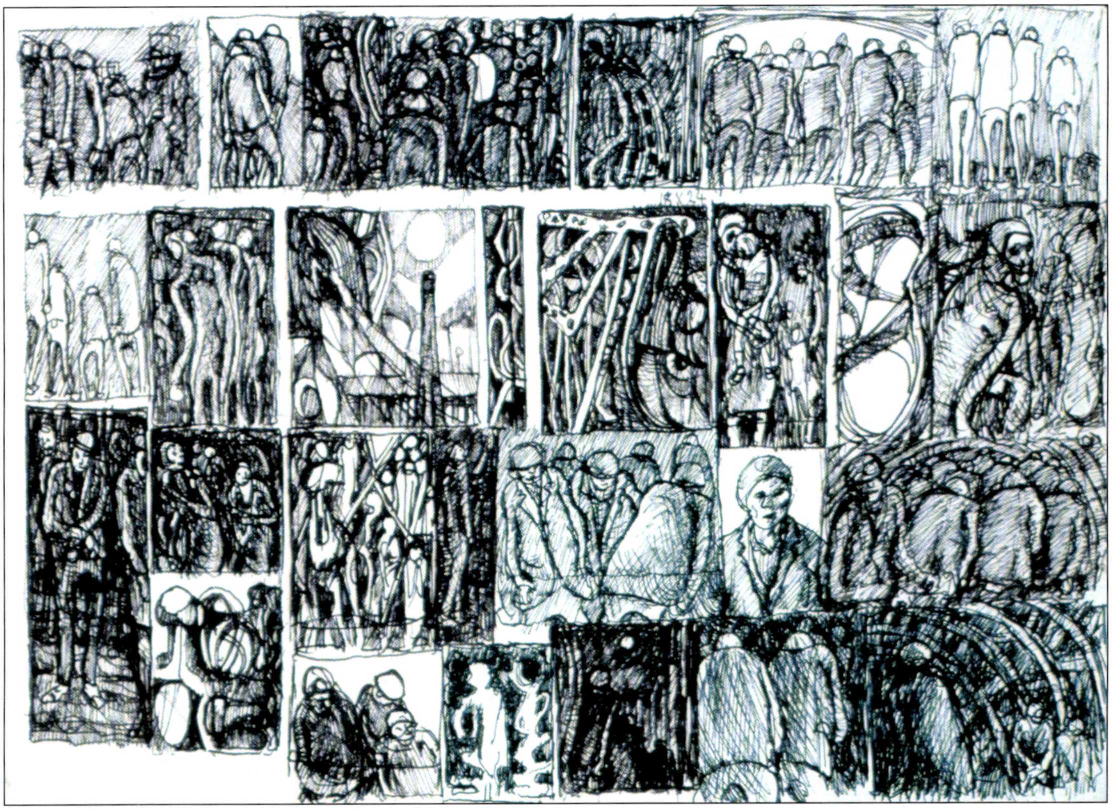

Plate 59

Weardale Steel Works 2001

Ink and mixed media - 50 x 35cm

Always fascinated by machinery and industrial processes, Tom was delighted to be asked to visit the foundry and observe the casting process at first hand. He videoed aspects of particular interest to ensure accuracy in the wash-offs and etchings which he subsequently produced. The semi-abstract wash-offs form a link between Tom's mining images and his more abstract wash-offs produced at Hartlepool Nuclear Power Station.

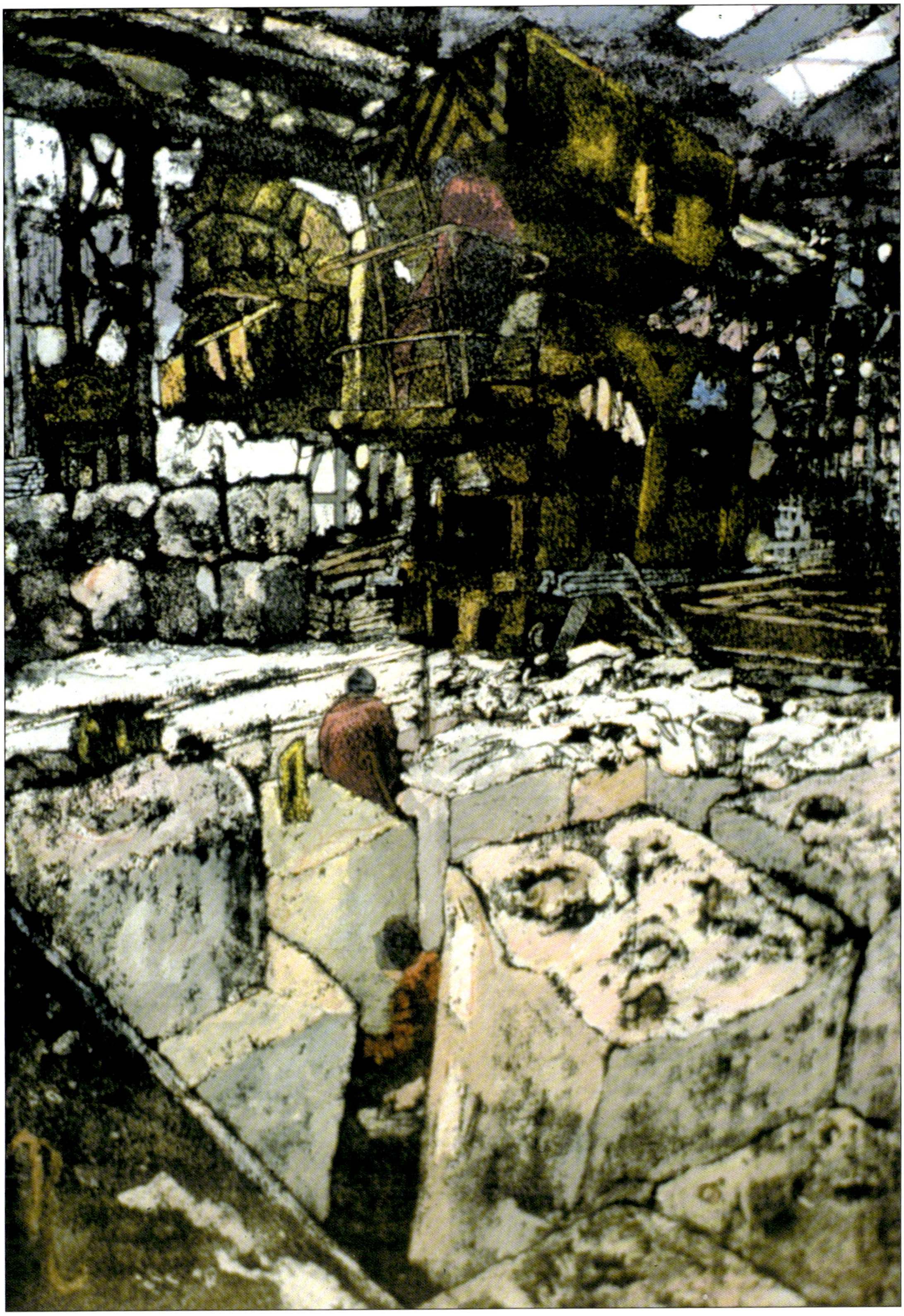

Plate 60

The Inspection Team 2006

Crayon on paper - 80 x 60cm

An imaginary interpretation of a team inspecting a disused tunnel, Tom conceded that the team would consist of fewer men but used artistic licence to produce a more-pleasing design. Drawn in monochrome with crayon, the roof has become the focal point for a human puzzle in which it is extremely difficult to identify all the characters in the maze of woodwork. This fascinating study took an entire week to produce such was the intricacy of the design, Tom grumbled!

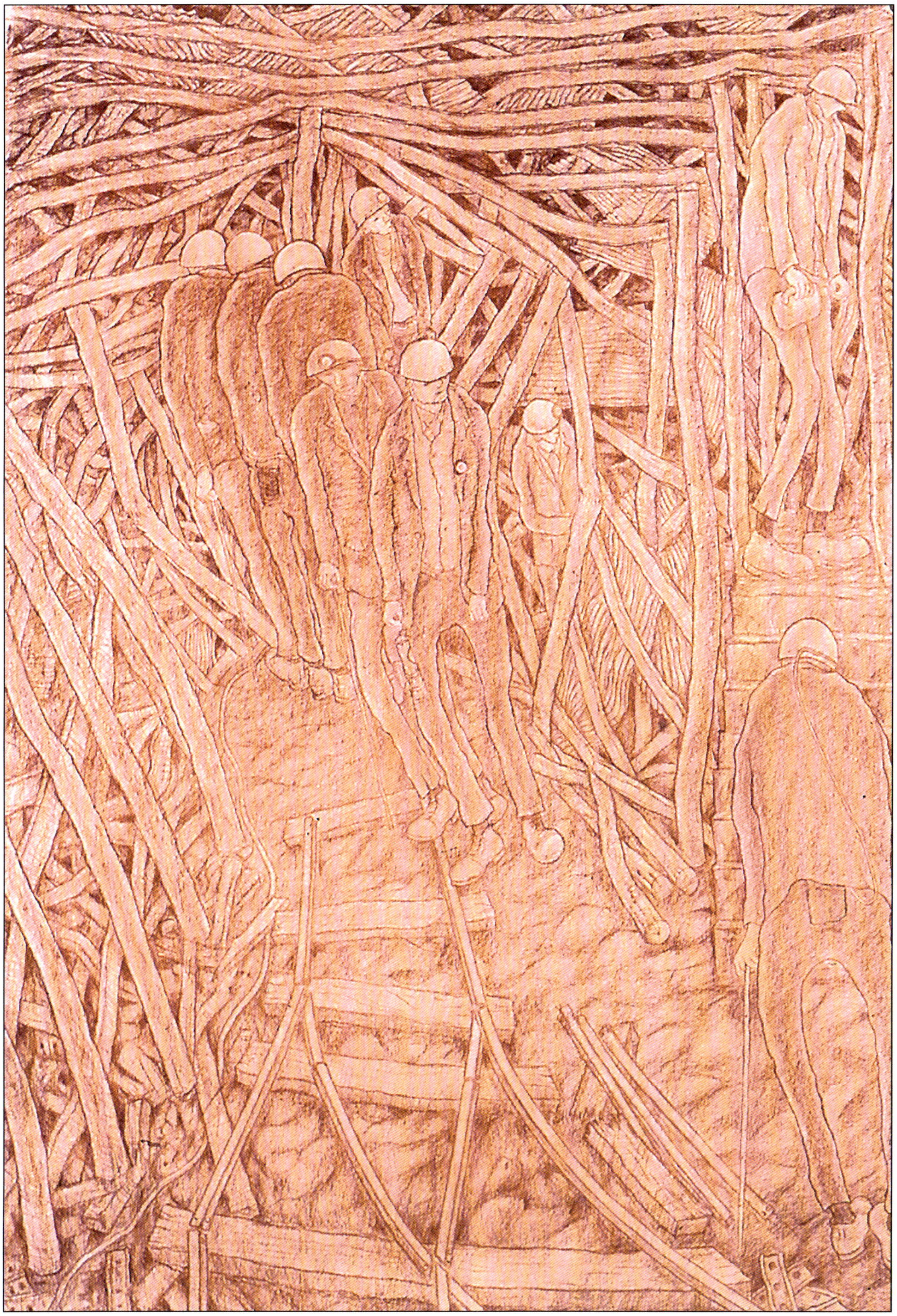

Tom McGuinness
30th April 1926 - 8th February 2006

Sadly, Tom McGuinness died during the preparation of this book. It was written, with his full co-operation, to coincide with what would have been his 80th birthday on 30th April 2006, as a celebration of his art and his life's work. The book was to have been launched during a major retrospective exhibition at The McGuinness Gallery in Bishop Auckland Town Hall. With Tom's family's encouragement these celebrations will still go ahead. As a tribute to Tom the text remains unaltered.

Fittingly Tom was painting up until his death - his paintbox remained open on his kitchen table with the unfinished oil painting standing on his easel where he had left it the night before. Although unbeknown to us at the time, Tom had asked us to include what turned out to be his final completed work. Appropriately this is the last plate in the book, 'The Inspection Team'.

For a modest, unassuming man of few words Tom has left behind a tremendous personal and artistic legacy. His priorities were always his family and his art. His loyalty lay with the mining community of South West Durham which features so prominently and perceptively in his work. A gentle man, it is through his art that Tom will live on and will never be forgotten. Great art speaks to all generations. Tom was a great artist - and a great friend.

Robert McManners and Gillian Wales - 15th February 2006

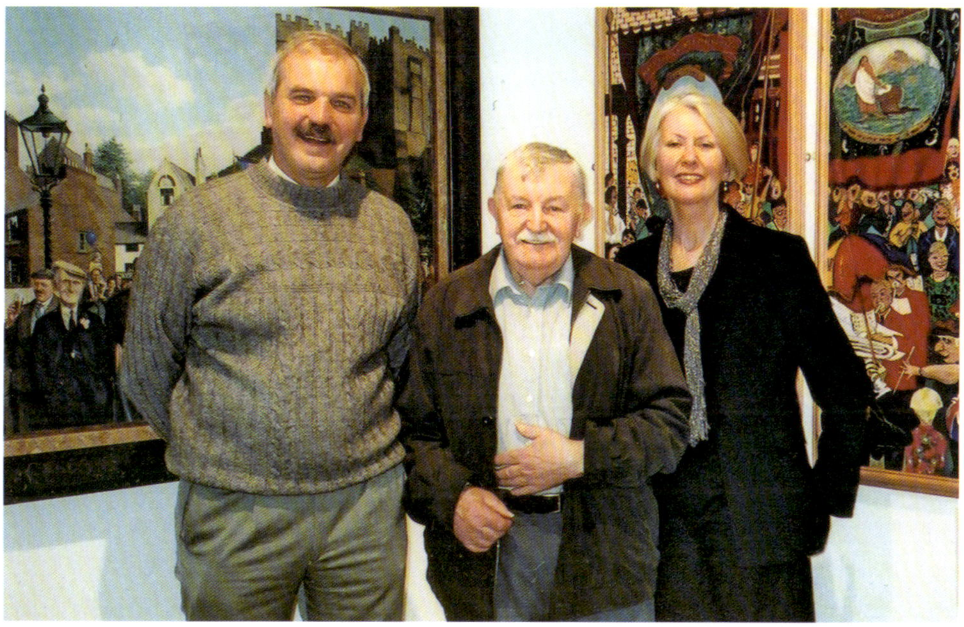

List of Plates

Selected Exhibitions

Group Exhibitions

1949 - Federation of Northern Arts Societies, Shipley Art Gallery, Gateshead.

1949 - Artists of the Northern Counties, Laing Art Gallery, Newcastle upon Tyne.

1951 - Contemporary Artists of the Northern Counties, Shipley Art Gallery, Gateshead.

1959 - Extra-mural Studies, Science Buildings, Durham University.

1964 - Group Exhibition, Shipley Art Gallery, Gateshead.

1964 - Group Exhibition, Bondgate Gallery, Alnwick.

1967 - Loan Exhibition of Contemporary Art, Laing Art Gallery, Newcastle upon Tyne.

1971 - Figurative Arts North, Laing Art Gallery, Newcastle upon Tyne.

1977 - Miner in Art, Grundy Gallery, Blackpool.

1978 - Art for Society, Whitechapel Art Gallery, London.

1978 - Life and Labour of the Miner, T.U.C. Headquarters.

1978 - Northern Arts Exhibition, Shipley Art Gallery, Gateshead.

1979 - Northern Print Workshops, Northgate Gallery, Newcastle upon Tyne.

1980 - Recent Prints from Lowick House, Abbot Hall Art Gallery, Kendal.

1981 - Northern Printmaking, Buddle Arts Centre, Wallsend upon Tyne.

1983 - Durham Miners' Gala Exhibition, Redhills, Durham City.

1983 - Coal - British Mining in Art, D.L.I. Museum and Art Gallery, Durham City.

1983 - Inaugural Exhibition, Miles Meehan Gallery, Darlington Arts Centre.

1985 - Tyne Tees Northern Open.

1987 - Second Northern Prints Exhibition, Newcastle upon Tyne.

1988 - A Celebration of the North, The Mall Gallery, London.

1988 - In Aid of Cheshire Homes, Wynyard Hall, Co. Durham.

1989 - Celebration of the North, Kirkleatham Old Hall Museum, Redcar.

1990 - Small Works Exhibition, Polytechnic, Newcastle upon Tyne.

1991 - Small Works Exhibition, Polytechnic, Newcastle upon Tyne.

1991 - In Aid of the N.S.P.C.C., Bowes Museum, Barnard Castle.

1991 - Children's Society, Arts Centre, Washington.

1992 - Children's Society, Arts Centre, Washington.

1992 - Northern Art Today 1992, St Oswald's Hospice, Gosforth.

1992 - A Fresh Look, Art Gallery, Middlesbrough.

1992 - Exhibition of Newcastle Artists, Central House of Artists, Moscow.

1993 - Children's Society, Arts Centre, Washington.

1993 - Our Working Heritage, Gossipgate Gallery, Alston.

1996 - Macmillan House Hospice, Hutchinson Gallery, Bishop Auckland.

1997 - The Miner, Biddick Arts Centre,Washington.
1998 - Underground and Overground, The School House, Carrshield, Allendale.
1998 - Federation of Northern Art Societies 50th Anniversary Exhibition, Laing Art Gallery, Newcastle.
2000 - Dark and Light, Broughton House Gallery, Cambridge.
2001 - Festival de L'Humanité, Paris.
2002 - Mining Art Exhibition, Lille.
2003 - Tom McGuinness and the Art of Industry, Ormesby Hall, Middlesbrough.
2003 - Exposition, McGuinness Gallery, Bishop Auckland.
2004 - Shafts of Light: Mining Art in the Great Northern Coalfield, Customs House, South Shields.
2005 - Coal Faces, Mining Lives, NCMME, Overton, Wakefield.
2005 - Face to Face: British Self Portraits in the Twentieth Century, UNN Gallery, Newcastle.

Solo Exhibitions

!957 - Solo Exhibition, Norman Godsmark Gallery, Darlington.
1958 - Coal Industry Social Welfare Organisation, Hobart House, London.
1960 - Solo Exhibition, Godsmark and Miller-Williams Gallery, Darlington.
1963 - Solo Exhibition, Swarthmore Arts Centre, Leeds.
1964 - Solo Exhibition, Art Gallery, Middlesbrough.
1965 - Solo Exhibition, I.C.A., Billingham.
1966 - Solo Exhibition, City Hotel, Durham City.
1972 - Solo Exhibition, John Whibley Gallery, London.
1977 - Solo Exhibition, Brampton Gallery, London.
1978 - Solo Exhibition, D.L.I. Arts Centre, Durham City.
1978 - Solo Exhibition, Grundy Gallery, Blackpool.
1978 - Solo Exhibition, Art Gallery, Middlesbrough.
1978 - Solo Exhibition, Library, Stockton-on-Tees.
1978 - Solo Exhibition, Markinch Centre, Fife.
1978 - Solo Exhibition, Sports Centre, Consett.
1978 - Solo Exhibition, Washington Arts Centre, Tyne and Wear.
1985 - The Power and the Glory, Dovecot Arts Centre, Stockton-on-Tees.
1985 - Solo Exhibition, Buddle Arts Centre, Wallsend.
1986 - Solo Exhibition, Greys Art Gallery, Gosforth.
1986 - Solo Exhibition, Bracken Press, Scarborough.
1986 - Solo Exhibition, Saddler Gallery, Durham City.
1987 - Solo Exhibition, Saddler Gallery, Durham City.

1988 - Solo Exhibition, Ormsby Hall, Middlesbrough.
1988 - The Working Eye, Broughton House Gallery, Cambridge.
1989 - Solo Exhibition, Art Approach Gallery, Chepstow.
1991 - Solo Exhibition, Polytechnic Gallery, Newcastle.
1992 - Solo Exhibition, Woodhorn Church Gallery, Ashington.
1992 - Solo Exhibition, Arts Centre, Washington.
1993 - Solo Exhibition, Seaton House, Easington.
1993 - Solo Exhibition, Hutchinson Gallery, Bishop Auckland.
1994 - A Way of Life, Broughton House Gallery, Cambridge.
1996 - The Jack Reading Bequest, Hutchinson Gallery, Bishop Auckland.
1996 - Backshift, The Mall Gallery, Crook.
1997 - Going Underground, Wellfield Court Gallery, Cardiff.
1997 - Mines a McGuinness, John and Josephine Bowes Museum, Barnard Castle.
1997 - The Jack Reading Bequest, Hutchinson Gallery, Bishop Auckland.
1998 - Mines a McGuinness, Bede's World, Jarrow.
1998 - McGuinness is Good for You, Hutchinson Gallery Bishop Auckland.
1998 - Mines a McGuinness, Doncaster Museum and Art Gallery, Doncaster.
1998 - Mines a McGuinness, NCMME, Overton, Wakefield.
1998 - Tom McGuinness, Scottish Mining Museum, Lady Victoria Colliery, Newtongrange, Midlothian.
1999 - Mines a McGuinness, Woodhorn Colliery Museum, Ashington.
1999 - Mines a McGuinness, Lancashire Mining Museum, Salford, Lancashire.
2000 - Mines a McGuinness, The Beacon Gallery, Whitehaven.
2001 - The Power and the Glory, Hutchinson Gallery, Bishop Auckland.
2001 - The Power and the Glory, Stanley Civic Hall, Stanley, Co. Durham.
2002 - Mines a McGuinness, Stanley Civic Hall.
2002 - The Shaft, McGuinness Gallery, Bishop Auckland.
2004 - Tom McGuinness and Friends, Gallerina, Darlington.
2004 - Face to Face, McGuinness Gallery, Bishop Auckland.
2005 - Etched in Coal, McGuinness Gallery, Bishop Auckland.
2005 - Mines a McGuinness, Art Gallery, Rugby.
2006 - McGuinness, McGuinness Gallery, Bishop Auckland.
2006 - McGuinness, Nunnington Hall, Helmsley, Yorks.

Glossary - Art Terms

Alla prima	A method of painting in a single layer on a white ground with no underpainting.
Aquatint	Specialised etching technique where the plate is coated with a porous resin to create a granulated effect.
Burin	An engraving tool.
Burr	The ridge of waste material raised by the burin when a metal plate is engraved.
Conté	Trade name for waxy chalks available in black, sepia and sanguine.
Distortion	Deliberate artistic device changing line, colour, tone, composition and perspective, often to heighten emotion.
Drypoint	Engraving directly on to a plate using a sharp needle, used to strengthen detail in an etching.
Etching	Design on a metal printing plate created by the action of acid.
Engraving	Printed image produced by cutting into the surface of a metal plate with a hardened instrument.
Expressionism	Art in which the emotion is more important than the observed reality, often using distortion.
Glaze	Transparent layer of paint applied over another of a different colour.
Gouache	Opaque watercolour.
Ground	A surface specially prepared for painting, perhaps with gesso or a layer of paint.
Iconography	The subject matter as opposed to the style of the image.
Iconology	Symbolism - the meaning underlying the image.
Intaglio	Printing where the surface of the print plate is incised to hold the ink which is then wiped and ink in the incised lines is transferred on to paper.
Lithograph	A drawing made on a stone or zinc plate using a waxy material which is then transferred to a print by applying a greasy ink.
Social realism	Art which realistically presents subjects of social concern.
Symbolism	Aims to suggest ideas by means of ambiguous yet powerful symbols.
Tempera	An emulsion used as a medium for a pigment traditionally made with egg yolk.

Glossary- Mining Terms

Backshift	The second shift - this varied from pit to pit between 6am and 9am starting time
Bait	Food taken to eat at work
Busty	The second lowest coal seam, worked extensively in Co. Durham
Canch	Part of the roof of the tunnel which is removed to make height. It is secured by arch-girders and lagging. The men who work it are known as Canch-men or Stone-men.
Cutter men	Men who operate the machine which literally cuts the coal at the coal face.
Deputy	Responsible for placing the men in the pit and for their safety.
Drawer	Miner who removes supports from the coal face as the face advances.
Foreshift	The first shift - times varied, starting any time from midnight to 6am.
Goaf	The empty unsupported space where the coal has been extracted.
Harvey	The fourth lowest seam worked extensively in Co. Durham.
Harvey Manrider	Conveyor belt used to transport men from one part of the pit to another.
Hewer	The miner who, using a hand pick, dug out coal and filled the coal tubs.
Inbye	In the direction of the face - away from the shaft.
Manrider	Underground passenger train used to transfer men to and from the coal face.
Mothergate	The underground roadway leading into the coal face.
Nightshift	The final shift of the day - time varied from colliery to colliery.
Onsetter	The miner responsible for the cage at the shaft bottom.
Outbye	In the direction of the shaft - away from the face.
Putter	Originally a young boy who brought empty tubs to the coal face for the hewer to fill. The putter pushed the full tubs to the shaft bottom either by his own power or with the aid of a pony.
Setter	Miner who erects vertical props at the coal face.
Tailgate	The underground roadway leading from the coal face.

Footnote - The Jack Reading Bequest is a collection of 50 works by Tom McGuinness. This collection is permanently held at Bishop Auckland Town Hall where, under the terms of the Bequest, it is displayed annually. These pictures represent the lifetime collection of the late Jack Reading.

Bibliography

Bindman, David ed. *Dictionary of British Art* Thames and Hudson 1995

Coal: British Mining in Art 1680 - 1980 Arts Council and NCB 1982

Davies, Alan *Facing Coal* Salford Museums Service 2001

Farrell, William *Selected Papers* Durham University Library collection, Palace Green

Freeland, Cynthia *But Is It Art?* O.U.P. 2001

Gombrich, E.H. *Art and Illusion; a study in the psychology of pictorial representation* Phaidon Press 1977

Gombrich, E.H. *The Story of Art* Phaidon Press 1984

Hall, Marshall *Artists of Northumbria* Art Dictionaries Ltd 2005

Holloway, Gill *A Bevin Boy Remembers* Henge Publications 1993

Lucy-Smith, Edward *Dictionary of Art Terms* Thames and Hudson 1993

Lucy-Smith, Edward *Symbolist Art* Thames and Hudson 2001

McManners Robert and Gillian Wales *Tom McGuinness: The art of an underground miner* Gemini Productions 1997

McManners, Robert and Gillian Wales *Shafts of Light: Mining Art in The Great Northern Coalfield* Gemini Productions 2002

Rosenberg, Jakob *Rembrandt: Life and Work* Phaidon 1968

Townsend, Imogen *Coal Faces Mining Lives* NCMME 2005

Vann, Phillip *Face to Face* Sansom 2004